Vadim M. Schastlivtsev, Vitaly I. Zel'dovich
Physical Metallurgy

Vadim M. Schastlivtsev, Vitaly I. Zel'dovich

Physical Metallurgy

Metals, Alloys, Phase Transformations

DE GRUYTER

Authors
Vadim M. Schastivtsev,
Ph.D and D.habil.,
Head Research Scientist, Laboratory of Physical Metallurgy,
Institute of Metal Physics, Ekaterinburg, Russia

Vitaly I. Zel'dovich,
Ph.D and D.habil.,
Head Research Scientist, Laboratory of Physical Metallurgy,
Institute of Metal Physics, Ekaterinburg, Russia

ISBN 978-3-11-075801-6
e-ISBN (PDF) 978-3-11-075802-3
e-ISBN (EPUB) 978-3-11-075812-2

Library of Congress Control Number: 2021949856

Bibliographic information published by the Deutsche Nationalbibliothek
The Deutsche Nationalbibliothek lists this publication in the Deutsche Nationalbibliografie;
detailed bibliographic data are available on the Internet at http://dnb.dnb.de.

© 2022 Walter de Gruyter GmbH, Berlin/Boston
Cover image: © Vadim M. Schastlivtsev, Vitaly I. Zel'dovich
Typesetting: Integra Software Services Pvt. Ltd.
Printing and binding: CPI books GmbH, Leck

www.degruyter.com

About the authors

Schastlivtsev, Vadim M., 55 years of experience in the field of metal science and heat treatment of steels, head of the Ural School of Metal Science. The main direction of scientific activity – phase and structural transformations in steels, mechanical properties of steels. Full member of the Russian Academy of Sciences, Ph.D and D.habil. in Metallurgical Engineering, Professor, chief researcher, laboratory of physical metallurgy, Institute of Metals Physics of Ural Branch of the Russian Academy of Sciences. Author of more than 350 scientific papers, including 10 monographs. The main works by V.Schastlivtsev are on martensitic, bainite, and pearlite transformations in steels, structural heredity, and steel single crystals.

Zel'dovich, Vitaly I., 50 years of experience in the field of physical metal science. The main area of scientific activity is phase transformations in metals and alloys. Doctor of physics-mathematical Sciences, Professor, chief researcher, laboratory of physical metallurgy, Institute of Metals Physics of Ural Branch of the Russian Academy of Sciences. Author of more than 250 scientific papers. The main works by V.Zeldovich are on the martensitic transformation and formation of austenite in iron alloys, on alloys with shape memory, on the impact of shock waves on metal materials.

The book was published in Russian in 2015
Translated into English by Yu.G.Gorelykh

https://doi.org/10.1515/9783110758023-202

Abstract

The fundamental tenets and actual data on phase equilibria and phase transformations in metals and alloys are presented. The book discusses phase equilibria in detail: numerous binary phase diagrams and basic state ternary phase diagrams. Perceptions about temperature–pressure diagrams and metastable diagrams are given. The processes of crystallization and related phenomena, diffusion and its mechanisms (theory and experiment, Kirkendall effect, etc.), plastic deformation, including large deformations, polygonization, and recrystallization are considered. A large section is devoted to phase transformations in the solid state (martensitic transformations, decomposition of supersaturated solid solutions, and ordering). A separate chapter deals with the diverse transformations in steels and iron alloys (pearlite, martensitic, and bainitic transformations, tempering of hardened steel, and austenite formation). Almost all the sections exemplify transformations and structures of real alloys.

The main views on these issues are concisely and exactly summarized. Some interpretations of the phenomena are still controversial; in this case, the most common point of view is expressed.

The book is intended for advanced undergraduate students, graduate students, and researchers. It can be useful to readers for a general acquaintance with the theory of phase equilibria and phase transformations.

https://doi.org/10.1515/9783110758023-203

Abstract

Foreword

The monograph is a short introductory course on physical metallurgy and is intended for the first acquaintance with the foundations of physical metallurgy. As follows from the table of contents, it is mainly devoted to phase equilibria and phase transformations. It deals with the following issues: phase diagrams, crystallization, diffusion in metals, plastic deformation, recrystallization, and phase transformations in solid state (martensitic transformations, decomposition of supersaturated solid solutions, and ordering). A separate chapter offers more detailed attention to phase and structural transformations in iron alloys and steels.

In writing the book, we have leaned on monographs and textbooks listed below. The history of the development of physical metallurgy (Section 1.2) has been written using a review of a three-volume edition, ed. R.W. Cahn and P. Haasen (1987, second edition). Textbooks compiled by S.S. Steinberg (1961), I.I. Novikov, G.B. Stroganov, and A.I. Novikov (1994) underlie the presentation of Chapter 2. Chapter 3 relies on using a textbook by S.S. Steinberg (1961) and a chapter taken from a three-volume edition, ed. R. Cahn (1968, first edition) and written by W.A. Tiller. Chapter 4 is based on the books of B.S. Bokshtein (1978) and P.G. Shewmon (1966). Chapter 5 comes from several books: R. Honeycombe (1972), S.S. Gorelik (1978, 2005), V.V. Rybin (1986), and so on. Chapter 6 contains three large sections. Section 6.2 is based on a book by G.V. Kurdyumov, L.M. Utevsky, and R.I. Entin (1977) and a chapter written by J.W. Christian, from a three-volume edition, ed. R. Cahn (1968, first edition). Section 6.3 utilizes a textbook by Ya. S. Umansky and Yu. A. Skakov (1978). A book by M.A. Krivoglaz and A.A. Smirnov (1958) is behind Section 6.4. A monograph by G.V. Kurdyumov, L.M. Utevsky, and R.I. Entin (1977) and a textbook by M.A. Smirnov, V.M. Schastlivtsev, and L.G. Zhuravlev (1999) are the key to Chapter 7. A complete bibliographic description of the cited sources is given at the end of the book (see Bibliography).

A distinctive feature of this book is the brevity of the text, including a large amount of illustrative materials. Graphs and images of microstructures help better understand and remember basic facts and statements of physical metallurgy. The issues presented in the book imply knowledge of the basics of thermodynamics and crystallography, as well as the theory of crystal structure defects (lattice defects).

The monograph is intended for senior students, graduate students, and researchers in physics who desire to master the "Metallovedenie" and "Heat treatment of metals" courses. It can be useful to readers for a general acquaintance with the theory of phase equilibria and phase transformations. For a more detailed study of the issues outlined, the readers can resort to the literature given at the end of the book. The course in physical foundations of metallurgy requires preparing at least the two following books:

https://doi.org/10.1515/9783110758023-204

1. Defects of the crystal structure and physical and mechanical properties of metals and alloys
2. Diffraction research methods (X-ray diffraction analysis, diffraction electron microscopy, and neutron diffraction)

The authors are grateful to A.E. Kheifetz, Ph.D. in physics and math, for helping in preparing the illustrative materials.

Contents

Chapter 1
History of the development of physical metallurgy

1.1 Physical and applied metallurgy

Through the centuries, man has accumulated knowledge about metals from everyday experience of their uses. Metals like iron, steel among others were always the major materials used for manufacturing weapons. Tools and kitchen utensils are another area. The Bronze Age changed the Iron Age, and the share of iron-based alloys and steels is currently 95% of all produced metals. However, various fields of technology apply alloys based on aluminum, titanium, copper, nickel, and other metals. Various combinations of the treatment of metals to obtain the desired products and their properties include deformation and heating. These operations lead to a change in the internal structure of metals and alloys. Physical metallurgy studies the internal transformations that occur during the treatment of metals, as well as the structure of the resulting metals and their properties. The ultimate goal of physical metallurgy as a science is to establish the treatment–transformation–structure–properties relationship. The practical task of physical metallurgy is to obtain metallic materials with the required performance characteristics. This is why physical metallurgy refers to an applied science.

Physical metallurgy has developed using ideas from other areas of knowledge. Mineralogy and crystallography have suggested that metals, like minerals, have an ordered structure. Thermodynamics and chemistry provide knowledge about equilibria and the kinetics of reactions. Metallurgy has offered information about the manufacture of metals. The development of physics have had a significant impact on the formation of physical metallurgy as an independent science, since metallurgists have used the concepts of solid state physics and applied physical methods of studying metals. Therefore, modern physical metallurgy can be called physical one.

1.2 The history of the development of physical metallurgy

Physical metallurgy as a science roots down to the study of the structure and properties of steel. D.K. Chernov (1839–1921) is credited to be the founder of scientific physical metallurgy. In the classic work "A Critical Survey of the Articles of Messrs Lavrov and Kalakutsky on Steel and Steel Guns and D. K. Chernov's Own Research on the Same Subject" (1868), Chernov established the existence of "critical" temperature points of steel. Point a corresponded to the temperature above which heating is necessary to bring about hardening and to obtain high hardness. D.K. Chernov found that this temperature for steel of a certain (eutectoid) composition is ranged within 700–750 °C. Point b matched the temperature above which steel accrues a fine-grained structure.

https://doi.org/10.1515/9783110758023-001

Point c was a melting point. Later, D.K. Chernov discovered that point d (~200 °C) stands for the beginning of the martensitic transformation. The discovery of the critical points gave impetus for plotting Fe–C diagrams. Thus, empirical research of steels acquired a scientific basis.

The works of P.P. Anosov and H.K. Sorby played an important role in studying the structure of steel. P.P. Anosov was the first to employ a microscope to examine the structure of metals (1831), and soon (1841) he discovered the secret of Damascus steel. H.K. Sorby, as a mineralogist, studied the structure of steel's specially prepared surfaces (thin sections) under a microscope (1864, 1887) and was the first who described the structure later called pearlite ("pearl component"). Then Sorby revealed a structure that was named martensite after the German metallographer Adolf Martens. Martens also used a microscope to take a close look at kinks, cracks, and crystallization processes (1878). He was among the first who pointed out to the relation between microstructures and mechanical properties.

Chernov's works were continued by Floris Osmond (1887), a French scientist and engineer. In his search for the critical points of steels during both cooling and heating, he suggested an allotropic theory of transformations in steel. He designated the transformation temperatures as A_{c1} and A_{c3} upon heating and A_{r1} and A_{r3} upon cooling. He also introduced the Greek letters α, β, γ, and δ for the iron phases. Thus, experimental data were obtained for constructing Fe–C diagrams.

An attempt to plot the Fe–C diagram was made by Albert Sauveur (1896), who used Osmond's results. His diagram was extremely sparse and very rough. Many researchers consider the Roberts-Austen diagram (1897) to be the first Fe–C diagram. Note that Roberts-Austen even earlier (1875) published a diagram of Ag–Cu alloys in temperature-chemical composition coordinates. Roberts-Austen clarified the nature of a solid solution of the high-temperature phase of iron, later named after him austenite. He was a versatile scientist. He was engaged in diffusion issues, investigated the influence of small additions of impurities on the properties of metals. After Roberts-Austen, in 1900, H.-W.B. Roozeboom published a Fe–Fe$_3$C phase diagram that accounted for the Gibbs phase rule. Roozeboom's diagram is close to modern and considered classic. He entered the names of the lines of the beginning and end of solidification of liquid metal: liquidus and solidus. Then, C.T. Heycock and F.H. Neville produced Cu–Sn diagrams, and thus the foundations of the theory of state diagrams of various systems were laid.

An important stage in examining the structure of alloys was the institution of the rule of phases by D.W. Gibbs (1876). Gibbs introduced the concepts of "phase" and "component." At first, nobody noticed this work of Gibbs, in particular, because it seemed to be a consequence of the general laws of thermodynamics. However, later Gibbs's theory of phase equilibria began to be widely used in many fields of knowledge: physics, chemistry, metallurgy.

The advances in crystallography had an influence on physical metallurgy. Auguste Bravais (1849) established the possibility of the existence of 7 syngonies and

14 translation lattices. E.S. Fedorov and Arthur Moritz Schoenflies (1891) independently put forward the theory of an ideal crystal lattice and discovered 230 spatial Fedorov's groups or the maximum possible number of ways of arranging atoms in three-dimensional space. The next century brought a complete experimental confirmation of this theory through the methods of X-ray structural analysis.

At the beginning of the twentieth century, Gustav Tammann intensively developed physical metallurgy as a science. He plotted a large number of phase diagrams but, unfortunately, inaccurate for today's purposes. G. Tammann investigated the mechanism of phase transformations. Looking into the solidification of organic crystals and metals, he pointed to the existence of two processes: the formation of crystallization centers and their growth. Later, this research was theoretically and experimentally developed by Volmer (1926), who laid the foundations for the modern theory of nucleation.

There are (were) many papers on studying in detail the kinetics of transformations in steels, which are aimed at constructing decomposition diagrams for supercooled austenite at isothermal exposures. The decomposition diagrams consist of curves of the beginning and end (often degrees) of transformation in temperature–time coordinates. Therefore, they are called temperature–time–transformation (TTT) diagrams but have also another name, C-curves, due to their shape. In the 1920s and 1930s of the last century, constructing such diagrams for steels of various chemical compositions was a favorite engage of scientists. Afterward, many scientists of different countries (among them are Albert Portevin, Davenport, Bain, Wefer, Lewis, S.S. Steinberg and V.D. Sadovsky) obtained reference data on such diagrams. Isothermal decomposition diagrams for supercooled austenite made it possible to determine the temperature–time heat treatment regimes required for producing the desired structure and properties of specific steels.

Since the 1920s, X-ray diffraction analysis has become the main method for studying the atomic-crystal structure of metals and alloys. If earlier a microscope and a thermocouple were the main tools of scientists, then X-ray diffraction (Laue, 1912) opened up fundamentally new opportunities in exploring metals and literally revolutionized physical metallurgy (as well as in other natural sciences). Arne Westgren and Robert P. Fragman (1922) established that the α-, β-, and δ-phases of iron have a body-centered cubic crystal lattice, whereas the γ-phase has a face-centered cubic one. W.L. Fink and E.D. Campbell (1929) determined that the martensite lattice in steel is body-centered tetragonal. Bain (1926) proposed an atomic-crystallographic mechanism of martensitic transformation, which is still the basis for describing this class of transformations. G.V. Kurdyumov and G. Sachs (1930) revealed the orientation relationship of the crystal lattices of martensite and austenite and obtained the orientation relations named after them. Later (1940, 1946), A.B. Greninger and Alexander R. Troiano measured a macroscopic shift during the martensitic transformation and showed that the habit plane of martensite crystals has large indices. Their research led to further construction of a phenomenological theory of martensitic transformations (1953).

The aging process of alloys has been discovered by accident. Wilm (1906) had left the quenched (quickly cooled) aluminum alloy for a few days but then found that the hardness of the alloy increased. The reason for this was not clear. Many years later, Merika (1919) explained this phenomenon by relating the aging process to a decrease in the solubility of components with decreasing temperature. He suggested that the increase in hardness takes place through the decomposition of a supersaturated solid solution with the precipitation of the smallest particles. Merika's work provoked numerous research projects on aging because it became clear in which alloy systems this process should be expected. The precipitation of particles was discovered much later. This finding became possible after special methods of X-ray structural analysis and transmission electron microscopy came. Using lauegrams, Andre Guinier and George D. Preston (1937, 1938) independently of each other discovered the signs of the initial stages of the separation of particles from a solid solution. These precipitates were called the Guinier–Preston zones. The later stages of aging often form plate- or rod-shaped precipitates regularly oriented relative to the matrix. By analogy with the structures observed in meteorites, such precipitates are called Widmanstätt. In 1923, Belyaev described the formation of Widmanstätt ferrite in low-carbon steels as another example of the process of decomposition of a solid solution. The emergence of Widmanstätt structures during aging testified to the orientation relationship between the crystal lattices of the precipitated particles and the matrix. This statement has underlain the atomic-crystallographic theory of the decomposition of a supersaturated solid solution.

Studying the ordering process apparently began in 1923 once Bain discovered superlattices, although earlier N.S. Kurnakov and A.N. Zhemchuzhny (1908) utilized the method of electrical resistance to investigate reactions of this type. Transformations in solids when superlattices are formed are called "disorder–order" reactions (ordering). Bragg and Williams (1935) were the first to publish a paper on the ordering theory. To determine the degree of order and the kinetics of the ordering process, they took the interaction energy of atoms as a basis. Subsequently, many works were published on this topic. Approximately, at the same time, works were carried out on the theory of alloys. In 1926, William Hume-Rothery set of basic rules for forming solid solutions. His outstanding monograph indicated the conditions for the mutual solubility of two elements in the solid state.

In 1930–36, it became understood that the crystal lattice is not perfect. The concepts of point lattice defects were introduced by Ya.I. Frenkel (interstitial atoms) and Walter H. Schottky (vacancies). Afterward, it was proved that vacancies are equilibrium defects and their equilibrium concentration was measured. Point defects play an important role in diffusion and phase transformations. Linear lattice defects (dislocations) were postulated by M. Polanyi, Egon Orowan, and G.I. Taylor in 1934 independently of each other. The concept of dislocations upon plastic deformation made it possible to explain the colossal difference between the measured and theoretically calculated values of strength. Despite the fact that the dislocations explained many of the features of plastic deformation, their existence has been thrown into doubt for a long time until their direct observation by transmission electron microscopy. Later it became clear that the dislocations play an important role in the processes of polygonization and recrystallization.

At that time, investigating plastic deformation was accompanied by exploring the destruction. One of the most important works was published by David J. Griffiths (1920). He postulated the emergence of the smallest cracks to be stress concentrators and nuclei of destruction in glass subjected to a load. He pointed out that the formation of fracture surfaces must consume energy. This statement later became the basis of the theory he created.

In the 1940s, the outbreak of World War II made scientific research focus on military needs. The demand for alloys for various purposes caused the intensive development of physical metallurgy. The beginning of the use of atomic energy was an important event in the postwar years. The volume of knowledge dramatically increased in the field of elementary particle optics. As a result, qualitatively new advanced devices such as an ion projector, an electron-beam microanalyzer, and an electron microscope were designed. An ion projector made it possible to "see" atoms, vacancies, dislocations, and grain boundaries. An electron-beam microanalyzer enabled one to determine the chemical composition on an area of $1 \ \mu m^2$. An electron microscope as a device combining a microscope (with magnification up to 10^6 times) and an X-ray apparatus became a powerful tool for testing both metallographic and crystalline structure. High-resolution electron microscopy succeeded in observing the atomic planes of crystal lattices, in examining the structure of grain boundaries. The usage of these new devices undertook a second revolution in physical metallurgy.

It is difficult to list all the important results obtained during these years. From 1950 onward, B. Chalmers continued studies on crystallization. He studied the topography of the solid-melt separation boundaries (interfaces) and dealt with the problem of dendrite formation. David Turnbull showed the importance of heterogeneous nucleus formation (1958). The doctrine of the structure of grain boundaries has received significant development. Earlier, Bragg (1940) and Burgers (1940) described a simple tilt boundary as a boundary that consists of regularly spaced edge dislocations. Such a boundary structure becomes more complicated by increasing the grain misorientation angle and a twist added. High-resolution electron microscopy provides important information about the structure of boundaries. A lot of research works and conferences on grain boundaries have been held, but the problem currently remains relevant, especially in connection with the production of nanomaterials.

The special attention of scientists was drawn to the diffusion mechanism. Ernest Oliver Kirkendall (1947) had carried out an original experiment that revealed the vacancy mechanism of diffusion of substitution atoms. Lawrence Stamper Darken (1948) run an analysis of the Kirkendall effect and calculated the diffusion coefficients of copper and zinc for this experiment. He also proved that the chemical potential gradient rather than the concentration gradient is the driving force behind the diffusion. The diffusion mechanism of interstitial atoms does not involve vacancies – interstitial atoms diffuse across interstitial sites. Many research works include the themes on diffusion along accelerated diffusion paths (along grain boundaries, along a free surface, along dislocations). The diffusion rate as a function of the grain misorientation angle along the grain boundaries was measured.

The use of transmission electron microscopy in the second half of the twentieth century gave a powerful impetus to the study of plastic deformation. Many works were aimed at establishing the relationship between the stages of deformation (a stress–strain curve) and the emerging dislocation structure. This relationship was established for metals with a face-centered cubic lattice (three stages of the hardening curve) and not so definitively for metals with other crystal structures. Studies of the yield strength led to the concept of Cottrell atmospheres – microregions of dislocations pinning by clouds of impurity atoms (e.g., dislocation pinning in iron by carbon atoms). Analyzing the relationship between strength characteristics and structure parameters showed that the strength can be enhanced by reducing the grain size. Hall–Petch's law (1953) reads that the yield strength or fracture stress is grain-size dependent. The theory of destruction was and remains extremely important in the field of applied physical metallurgy. The dislocation mechanism of initiation of cracks and the mechanism of their propagation were proposed. Special attention was paid to the passage from ductile (plastic) fracture to brittle fracture with decreasing temperature. This is because brittle fracture provokes to catastrophic accidents.

The processes occurring during heating (annealing) of deformed metals have been the subject of numerous examinations. Modern physical metallurgy divides these processes into recovery (includes relaxation and polygonization) and recrystallization. The recovery phenomenon was studied by G. Tammann. S.T. Konobeevsky and I.I. Mirer (1932) were the first to observe polygonization in rock salt crystals. The mechanism of polygonization was explained later, with the development of the concept of dislocations. It is difficult to ascertain when and by whom recrystallization research began. It is believed that the first observation of grain growth upon heating a deformed metal (zinc) was carried out by Salomon Kalisher (1881). In 1898, John Edward Stead affirmed that recrystallization occurs by nucleation and growth. The nucleation mechanism has been the subject of lengthy discussions. To date, several models of forming recrystallization centers have been proposed. These models are based on the transformation of low-angle boundaries being in a deformed metal or arising during polygonization into high-angle boundaries capable of intensive migration. Regardless of the study of the recrystallization mechanism, Johnson and Meil, as well as Avrami, analyzed the formal kinetics of the process. The kinetics of recrystallization obeyed an equation previously proposed by A.N. Kolmogorov to describe the kinetics of crystallization. To approximately estimate the temperature of the onset of recrystallization, A.A. Bochvar proposed a rule through a ratio between this temperature and the melting point temperature. It should be emphasized that the recrystallization temperature is not constant, but depends on many factors.

One of the main trends of research in physical metallurgy was studying phase equilibria and phase transformations in solids. Larry Kaufman and G. Pettsov made an important contribution to techniques for calculating phase diagrams. Nevertheless, until now, an experiment remains the main way for constructing phase diagrams. Research works on the decomposition of supersaturated solid solutions was the starting point for putting forward the spinodal decomposition theory by M. Hillert and R.W. Cahn. In the 1940s, appropriate modulated microstructures were observed by Lipson and Daniel by the X-ray method, and subsequently (25 years later) by Nicholson and Taft by the method of transmission electron microscopy. Findings on ordering alloys obtained a better understanding through the use of neutron diffraction and electron microscopy. Electron microscopy gave information on the domain structure of ordered alloys (not to be confused with the domain structure of ferromagnets). There often arose a question: what mechanism of the ordering process is – either nucleation and growth of the ordered phase regions or without nuclei by a gradual increase in the degree of long-range ordering throughout the entire volume. It turned out that both mechanisms can be brought about but with different ordering types in alloys.

Papers on phase transformations in steels did not stop publishing. The variety of martensitic transformations (and not only in steels) posed new challenges for scientists and led to new results. Let us list the main ones. In the early 1960s, V.F. Zackey and Earl R. Parker proposed special steels (TRIP steels). The latter turned out to be remarkable by the enhancement in ductility characteristics (at high strength) due to a controlled martensitic transformation. A new class of steels (maraging carbon-free steels) was created in which high mechanical properties (strength, ductility, and fracture toughness) were achieved through a combination of martensitic transformation and aging. As a result of a deep understanding of the transformations that occur during heat treatment, another new class of steels was developed, ferrite-martensitic steels. In 1949, G.V.Kurdyumov and L.G. Khandros discovered thermoelastic martensitic transformations. Since the 1970s, this theme has caught the minds of many scientists, as it underlay new phenomena: shape-memory and superelasticity effects. In this context, nickel and titanium alloys near the equiatomic composition generated fresh interest due to their high mechanical and anticorrosive properties. Great merits in their research and practical application belong to C.M. Wayman, K. Otsuka, and V.A. Likhachev.

In 1960, P. Duwez came up with an invention of a rapid quenching process from a liquid state. At cooling rates of the order of a million degrees per second, alloys of a certain chemical composition could be obtained in an amorphous state. In other words, a metal melt solidified without crystallizing. One of the features of amorphous alloys is the lack of texture. Amorphous alloys have unusual properties and find some application. Another important result of rapid quenching is the formation of a fine-grained state of the alloy with increased strength characteristics.

The physical metallurgy of ferromagnetic materials developed at a rapid pace. Here is a non-exhaustive list of the names of scientists whose works have made a significant contribution to the theory of magnetism: Richard M. Bozorth, Louis Neel, Sergey V. Vonsovsky, Edmund C. Stoner, and Erich P. Wohlfarth. The nature of the high-coercive state in various alloys has been established, and materials for permanent magnets with extremely high magnetic energy values have been created. Many research works have been conducted to produce the most favorable texture and improve the magnetic properties of transformer steel.

It is of interest to track progress in the study of superconducting materials. It should be underscored that there are many Nobel Prize winners among scientists dealing with the problem of superconductivity. In 1911, the phenomenon of superconductivity on mercury was discovered by Heike Kamerlingh-Onnes. He revealed a critical temperature and a critical magnetic field that destroys superconductivity. In 1933, Walther Meissner found that an external magnetic field at strength below a critical value does not penetrate into a superconductor. In 1950, Vitaly L. Ginzburg and Lev D. Landau were the first who devised a phenomenological theory of superconductivity. In 1957, John Bardeen, Leon Cooper, and John Robert Schrieffer constructed a microscopic theory of superconductivity (BCS theory). In 1957, A.A. Abrikosov put forward the idea that magnetic flux can enter a superconductor in the form of vortices attached to structural defects. In 1986, J. Georg Bednorz and K. Alex Müller made a breakthrough in exploring superconductivity. They found high-temperature superconductivity (HTSC) and were awarded the Nobel Prize a year later. The critical temperature of a discovered superconductor was 30 K, exceeding the temperature of liquid hydrogen. An intensive study of HTSC followed. In 1993, a superconductor with a critical temperature of 135 K was fabricated at Moscow State University. At present, the use of HTSCs in practice is a stubborn task. Progress in this field involves solving complex technological challenges.

Russian scientists have made a significant contribution to the development of physical metallurgy. Apart from the names of scientists mentioned above, the surnames of S.T. Kishkin, who met great success in producing heat-resistant alloys, and I.N. Friedlander, who designed a number of aluminum alloys for aircraft construction, deserve special respect. Under Bochvar's charge, many non-ferrous alloys were explored, and materials for nuclear engineering were composed. V.D. Sadovsky and his colleagues carried out numerous fundamental studies of phase and structural transformations in steels. It is impossible to roll the names of all the scientists, by whose labors the basis of modern physical metallurgy was building. We apologize to those who were not mentioned in this brief overview.

This section does not discuss the history of the development of mechanical test methods and provides no information on a progressive enhancement of the mechanical properties of metals and alloys. This circumstance explains the fact that this monograph is devoted mainly to phase equilibria and phase and structural transformations. Strength and ductility issues, mechanical test methods should be covered in another course.

1.3 Course content

The book is a short introductory course on physical metallurgy and intended for the first acquaintance of students and scientists not previously majoring in the "Science of Metals" and "Heat Treatment of Metals" courses with the foundations of physical of metallurgy. We believe that the monograph can be also useful for evaluation purposes on the concept of phase equilibria, phase transformations, plastic deformation, and recrystallization.

The "Phase transformations in metals and alloys" lecture course delivered by one of the authors (V.I. Zeldovich) at the Faculty of Physics of the Ural State University named after A.M. Gorky underlies the content. The fundamental principles and actual data on phase equilibria and phase transformations in metals and alloys are given. The present course includes phase diagrams of metal systems, crystallization, diffusion in metals, plastic deformation, recrystallization, phase transformations in the solid state (martensitic transformations, decomposition of supersaturated solid solutions, ordering), and transformations in steels. An extremely broad field, physical metallurgy, overlaps with such areas of physics as crystallography, X-ray diffraction analysis, and electron microscopy – which topics go beyond the scope of this book. Also, the doctrine of crystal lattice defects and physical and mechanical properties of metals and alloys should be presented in a separate course.

Chapter 2
Phase diagrams

Phase diagrams play an important role in physical metallurgy. They serve as the basis for analyzing phase transformations and forming the structure of alloys when changing their chemical composition, temperature, and pressure. Let us first resort to definitions of the concepts used further in this chapter. *An alloy* is referred to as a mixture obtained by fusing two or more elements where at least one of them is metal. *Components* are called chemically individual substances combined into an alloy. Components can be both elements and chemical compounds. *A phase* is the homogeneous part (solid and liquid) of an alloy, isolated from another parts by a separation surface demarcating discontinuous change in alloy properties. In real alloys, the phase includes a set of homogeneous alloy parts having a separation surface (interphase boundaries). *Equilibrium* of the phases is reached when there are no visible transformations in an alloy. In other words and more precisely, an alloy is in a state of thermodynamic equilibrium if its state parameters (temperature, pressure, and concentration) do not change over time and if there are no fluxes of any type. From a microscopic point of view, such a state is a state of dynamic (mobile) equilibrium. *A phase diagram* involves a graphic representation of the equilibrium states of the phases in the system of a given group of alloys.

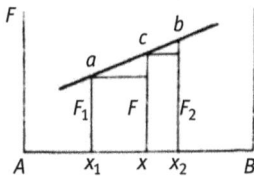

Figure 2.1: Free energy of a mechanical mixture of phases.

2.1 The phase rule

The conditions of equilibrium of the phases in an alloy are satisfied when chemical potentials of each component in any of the phases are said to be equal. If a given component in one phase has the chemical potential greater than in the other phase, it passes to a phase with lower chemical potential. The condition of minimum thermodynamic potential dictates the equality of the chemical potentials of each component in all phases being in equilibrium. Let k be the number of components and Φ be the number of phases.

Suppose that $\mu((j)/i)$ is the chemical potential of the ith component in the jth phase. Gibbs' phase rule provides the possible number of independent parameters (the number of degrees of freedom) of a closed system, with the system remaining

https://doi.org/10.1515/9783110758023-002

in equilibrium even if they are changed. If equilibrium is established in the system with k-components and Φ-phases, then the temperature T and pressure P are the same in all phases. The equilibrium conditions are written as follows:

$$\mu_1^1 = \mu_1^2 = \mu_1^3 = \cdots = \mu_1^\Phi,$$

$$\mu_1^1 = \mu_1^2 = \mu_1^3 = \cdots = \mu_1^\Phi,$$

$$\cdots\cdots\cdots\cdots\cdots\cdots\cdots\cdots,$$

$$\mu_\kappa^1 = \mu_\kappa^2 = \mu_\kappa^3 = \cdots = \mu_\kappa^\Phi.$$

These conditions can also be presented as $(\Phi - 1)\,\kappa$ equations. The number of variables in the system is

$$\Phi(\kappa - 1) + 2,$$

where $\Phi\,(\kappa - 1)$ is the number of component concentrations in all phases, and 2 is the temperature and pressure. The number of degrees of freedom, C, is obtained by subtracting the number of equations from the number of variables:

$$C = \Phi\,(\kappa - 1) + 2 - (\Phi - 1)\,\kappa = \kappa - \Phi + 2.$$

We have arrived at Gibbs' phase rule:

$$C = \kappa - \Phi + 2.$$

As the number of degrees of freedom cannot be negative: $C \geq 0$, the phase rule should be

$$\Phi \leq \kappa + 2.$$

2.2 Thermodynamic justification for constructing phase diagrams

In thermodynamics, the free energy F is written as follows:

$$F = H - TS,$$

where H is enthalpy, T is temperature, and S is entropy. A change in the free energy with changing temperature at constant pressure P amounts to

$$\left(\frac{\partial F}{\partial T}\right)_P = -S.$$

Entropy being always greater than zero ($S > 0$), the free energy always decreases as temperature increases. It is known from thermodynamics that the entropy of a solid solution is always greater than that of a mechanical mixture. The entropy of a fluid is always greater than the entropy of a solid. Therefore, when temperature rises, the free energy goes down more strongly for a liquid, weaker for a solid solution, and even weaker for a mechanical mixture. A change in temperature leads to a change in the mutual arrangement of the concentration dependencies of the free energy. For a given group of alloys, this allows determining the sequence of phase transformations during heating and cooling.

For mechanical mixtures and for solid solutions, consider how the free energy behaves in an $A-B$ binary system as their concentration changes. On the abscissa axis, we plot the concentration of a component, and the ordinate axis presents the free energy. The concentration axis is limited; it extends from zero to 100% (weight or atomic). The pressure is assumed to be constant. It is easy to show that the dependence of the free energy of a mechanical mixture of phases on the concentration has the form of a straight line. Let the point a be the free energy F_1 of the phase with concentration x_1 and the free energy F_2 of the phase with concentration x_2 be denoted by the point b. Then the free energy F of the mechanical mixture of these phases is represented by the point c residing on the ab segment (Figure 2.1). The dependence of the free energy of a solid or liquid solution on the concentration has the form of a sagging line (Figure 2.2). Since the free energy of a solid solution is always less than the free energy of a mechanical mixture, any point of the curve must lie below the straight line depicting the free energy of the mechanical mixture. For example, the point c lies below the segment ab (Figure 2.2). In the case of limited solubility, the dependence of the free energy on concentration has a more complicated form (Figure 2.3). The free energy of solutions on the ab segment is greater than that of mechanical mixtures. Therefore, solid solutions within this concentration range are unstable and must segregate into a mechanical mixture of solid solutions with the concentrations x_1 and x_2 (Figure 2.3).

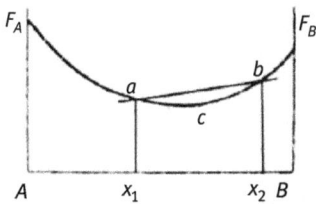

Figure 2.2: Free energy of a solution.

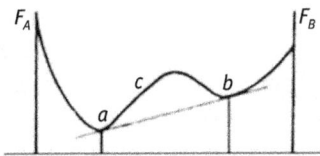

Figure 2.3: Limited solubility: free energy.

In this case, the point c lies above the ab segment. The straight line intersects the curve of the concentration dependence of the free energy of the solutions at points a and b. The latter correspond to the concentrations x_1 and x_2. Solid solutions are stable to the left of the point a and to the right of the point b.

Geometric thermodynamics claim that the equilibrium of a liquid solution and a crystallizing solid solution can be described by a segment of a common tangent to the free energy curves (Figure 2.4). Here, F_A and F_B are the free energies of the components in the liquid state; F_A' and F_B' are their free energies in the solid state; ab is a common tangent segment showing the concentrations x_1 and x_2 of the solid and liquid solutions in equilibrium, respectively.

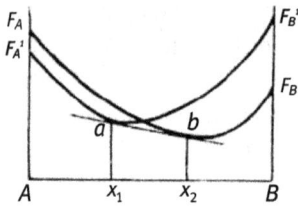

Figure 2.4: Equilibrium between liquid and solid solutions.

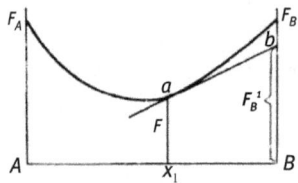

Figure 2.5: Equilibrium between a pure component and a solution.

The equilibrium between the liquid solution and crystals of the precipitated component B should be regarded as a segment tangential to the curve of the free energy for the liquid. The segment passes through the free energy point of the component B (Figure 2.5). Here F and F_B' are the free energies of the liquid solution and the solid component B in equilibrium; the ab segment indicates the concentrations of the coexisting phases.

The stated rules of geometric thermodynamics underlie constructing state diagrams for binary and more complicated systems.

2.3 Unlimited solubility diagram: the lever rule

To construct a binary phase diagram, we should plot the concentration of a component along the abscissa axis; the temperature should be located on the ordinate axis. As in the previous case, the concentration varies from zero to 100%. The pressure, if not specifically stated, is thought to be constant (atmospheric). The figurative point in the diagram indicates the state of an alloy of a given concentration at a given temperature. The point shows a change in temperature extended along the vertical axis and a change in concentration read off along the horizontal one. Let us first consider one of the simple diagrams, the phase diagram of unlimited solubility alloys in the liquid and solid states. An example of such a diagram is a Cu–Ni system.

The components A and B form a continuous series of solid solutions. For complete mutual solubility, three conditions are required. The components must have the same crystal lattice. The difference in the size of atoms should not exceed ~15%. The interaction between atoms of different types must be stronger than that between atoms of the same type.

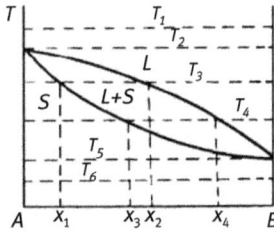

Figure 2.6: A full mutual solubility diagram for liquid and solid states.

Figure 2.6 displays the phase diagram of completely mutual soluble alloys in the liquid and solid states ("a cigar"). Above the liquidus line L, the alloys are in the liquid state; below the solidus line S, the alloys are in the solid state. These are single-phase areas of the diagram. Inside the cigar, liquid and solid solutions coexist; this is a two-phase region $(L + S)$. Figures 2.7–2.12 present the mutual arrangement of the free energy curves for the liquid $(F_A - F_B)$ and solid solutions $(F'_A - F'_B)$ at different temperatures. Changing this location dictates the crystallization process. When lowering the temperature, first, the refractory component A is crystallized (Figure 2.8). Then solid solutions rich in component A (Figure 2.9), and solid solutions rich in component B are crystallized (Figure 2.10). Finally, the low-melting component B is crystallized (Figure 2.11). In Figures 2.9 and 2.10, the position of the tangency points relative to the free energy curves is responsible for the concentrations (x_1, x_2, x_3, x_4) of solutions that are in equilibrium with each other. The phase diagram (Figure 2.6) reflects the position of the corresponding temperatures and concentrations of crystallizing solid solutions.

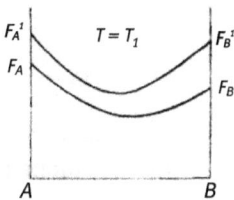

Figure 2.7: Stability of liquid solutions.

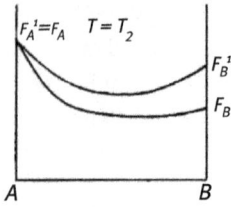
Figure 2.8: Crystallization of component A.

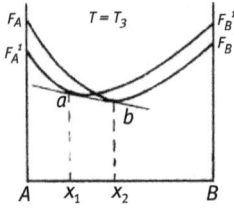
Figure 2.9: Crystallization of component A-rich solutions.

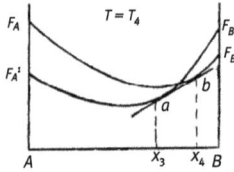
Figure 2.10: Crystallization of component B-rich solutions.

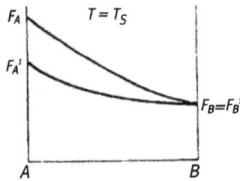
Figure 2.11: Crystallization of component B.

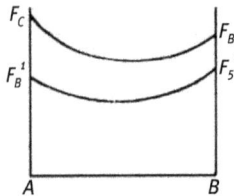
Figure 2.12: Stability of solid solutions.

Examples of unlimited solubility diagrams are listed in reference books on phase diagrams of binary alloys. Among the latter, the Cu–Ni system mentioned above, Bi–Sb, Ge–Si, and other alloys can be found there.

One of the unlimited solubility diagrams is a diagram with a minimum (Figure 2.13). This diagram is a result of decreasing the crystallization temperature of A- or B-component-based solutions with increasing concentration of the component B or A, respectively. The free energy curves for the liquid and solid solutions are mutually arranged within the crystallization temperature range, as in Figure 2.14. Alloy systems such as Au–Cu, Co–Pd, and Ti–Zr, have similar diagrams.

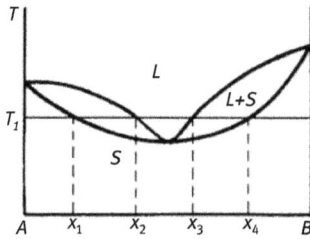

Figure 2.13: A full mutual solubility diagram with a minimum.

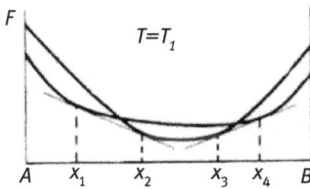

Figure 2.14: Arrangement of the free energy curves within crystallization temperatures.

The phase rule helps determine the number of phases in equilibrium. Since the pressure is assumed to be constant, the phase rule appears as follows:

$$\Phi \leq K + 1.$$

As it follows from the rule, an equilibrium two-component system can have not more than three phases, including the liquid phase. Then, applying the ratio

$$C = K - \Phi + 1,$$

we can calculate the number of degrees of freedom. The single-phase region of the binary diagram has two degrees of freedom because both the temperature and the concentration can be changed. There is only one degree of freedom in the two-phase region: the concentration of the phases changes when varying the temperature. In other words, the concentration is possible to change if the temperature varies.

Consider a rule for calculating the quantitative ratio of equilibrium phases in a two-phase region at a given temperature. First, the conoid concept should be introduced. A conoid is called a segment of a horizontal line connecting the compositions of coexisting equilibrium phases. Let x_1 be the percentage of component B in a liquid L at a given temperature, x_2 be the percentage of the component B in a solid solution S, and x be the average content of the component B in an alloy R (Figure 2.15a). The percentage of the liquid in the mixture $(L + S)$ may be denoted by p, and the percentage of the solid solution is q.

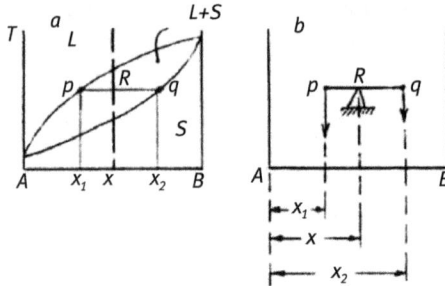

Figure 2.15: A diagram of the cigar type. A conoid pq (a), and the scheme for the lever rule (b).

We write down the system of equations as follows:

$$p + q = 1,$$

$$px_1 + qx_2 = x.$$

Having solved the system, we come up with the following results:

$$p = \frac{x_2 - x}{x_2 - x_1}; \quad q = \frac{x - x_1}{x_2 - x_1}; \quad \frac{p}{q} = \frac{x_2 - x}{x - x_1}.$$

The equation obtained is the lever rule (the rule of segments) for determining the quantitative ratio of coexisting phases. The graphic presentation of this rule is shown in Figure 2.15b, where an analogy with the lever rule of the first kind in mechanics can be traced.

2.4 A eutectic diagram

Another simple diagram is a eutectic diagram when the solubility of alloy components in the liquid state is unlimited and is almost absent in the solid state. Eutectic is an alloy either in the liquid or solid state; in the first case, the alloy is in equilibrium with crystals of the initial components and crystallizes at a constant temperature T_e (a eutectic point); in the second case, the alloy represents a mechanical mixture of crystals of the components.

Figure 2.16 presents a eutectic diagram.

Above the liquidus line L, alloys are in the single-phase liquid state, and other areas of the diagram are of two-phase. Figures 2.17–2.21 show the mutual arrangement of the free energy curves in the liquid $(F_A - F_B)$ and solid $(F'_A - F'_B)$ states as temperature drops. Reaching a temperature T_2, the refractory component B crystallizes (Figure 2.18).

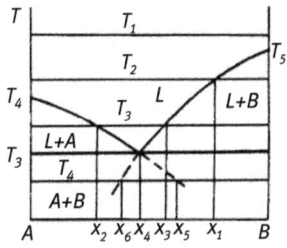

Figure 2.16: A eutectic diagram.

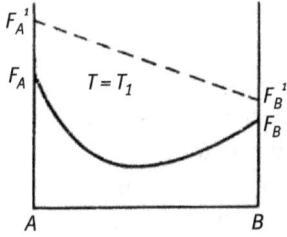

Figure 2.17: Liquid solution stability.

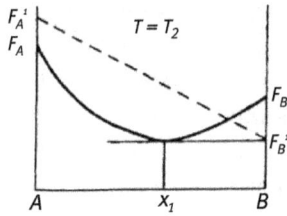

Figure 2.18: Crystallization of refractory component B.

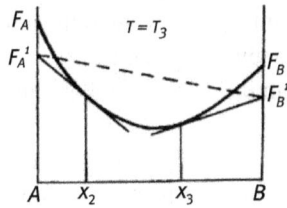

Figure 2.19: Crystallization of both components of different compositions.

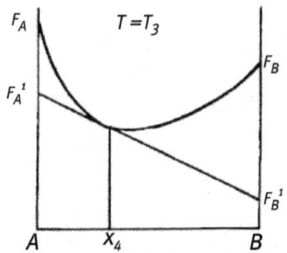

Figure 2.20: Crystallization of a eutectic.

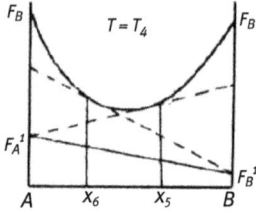

Figure 2.21: Solid phase stability.

The position of the tangency point x_1 speaks about equilibrium between a liquid and precipitated crystals of B. This point specifies the chemical composition of the liquid.

At a temperature T_3, both components in alloys of different compositions crystallize (Figure 2.19). The position of the tangency points x_2 and x_3 allows finding the composition of liquids in equilibrium with precipitated crystals of A and B. With a further decrease in temperature, the free energy straight line for the mechanical mixture of both components intersects the free energy line for the liquid (see Figure 2.20). At this temperature, the system of three phases, the liquid solution of composition x_4 and crystals of components A and B, is balanced. This temperature is called the eutectic temperature; the point x_4 is eutectic. At a temperature T_4, the mechanical mixture of crystals of both components is stable. At this temperature, however, it is possible to draw tangents to the free energy line of the liquid solution (dashed lines). The equilibria of the A and B components with the liquid solutions of the composition x_5 and x_6, respectively, are metastable. This occurs because their free energy is greater than the free energy of the mechanical mixture of components. The position of the corresponding temperatures and concentrations of the liquid solution in this system of alloys is shown in the diagram (Figure 2.16). The described concentration changes correspond to the states of the alloy when diffusion in both the liquid and solid phases leads to equilibrium at a given temperature.

Eutectic phase diagrams are quite numerous. For example, these are diagrams of Bi–Cd, Au–Si, Ag–Si, Al–Sn, and Be–Si alloys.

According to the rule of phases, it can be revealed that three phases (crystals of components A and B and a liquid) in a eutectic two-component system are in equilibrium at a eutectic temperature. In this case, the number of degrees of freedom is zero, and the system is said to be nonvariant. The ratio of the numbers of coexisting phases can be calculated by using the lever rule. For example, it is possible to determine the number of phases in an alloy with a concentration x at a certain temperature corresponding to the two-phase region ($L + B$) (Figure 2.22).

The amount of the liquid is equal to the ratio of the lengths of the segments BC/AC, and the number of crystals of B is AB/AC.

Apart from the concept of the phase component, the concept of a structural component should be entered. A part of a microstructure consisting of two or more phases and having the same structure is called a structural component. So, a mechanical mixture of phases, obtained by crystallizing the eutectic, and possessing,

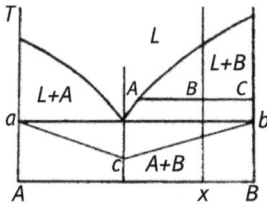

Figure 2.22: Applying the lever rule: Tamman's triangle method.

as a rule, a lamellar structure, can serve as an example of the structural component. Its number can be computed via Tamman's triangle abc (Figure 2.22). The measure of the area occupied by the structural component is the length of the vertical segment from the eutectic horizontal to the side of the triangle abc. The height of the Tamman triangle corresponds to 100% of the structural component area.

Consider the movement of a figurative point along a closed $abcd$ contour (Figure 2.23).

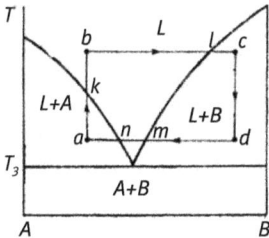

Figure 2.23: Transformations as the temperature and concentration change.

At the point a, an equilibrium alloy consists of crystals of component A and a liquid solution of composition n. As the temperature rises, the crystals of A dissolve and disappear at point k. As the concentration of component B increases (along the bc line at the point l), crystals of the component B appear in the liquid. The number of crystals of B increases and the amount of the liquid decreases when dropping the temperature (along the cd line). With a decline of the concentration of the component B (along the da line), the amount of liquid goes up, and the number of crystals of B falls down to zero at point m. There exists only a liquid in the mn segment. At point n, crystals of the component A emerge.

It should be emphasized that only one phase appears or disappears when crossing the line of the phase diagram. The above statement also extends to multicomponent phase diagrams and is called the law of "adjoining state spaces."

2.5 Experimental construction of phase diagrams

To theoretically construct phase diagrams, it is necessary to know the specific functional dependence of free energy on temperature T, pressure P, and concentration x.

Theoretically, there are various approximations of the theory of solutions (ideal, regular, etc.) for establishing this dependence. Experimentally, different physical and chemical methods of determining thermodynamic quantities are used to find it. The theory of regular solutions makes it possible to deduce equations of phase equilibrium curves and to meet criteria for the formation of phase diagrams of the simplest types. A further improvement in the theory of solutions and applying advanced computer technologies are expected to help take an active role in theoretical calculations for constructing phase diagrams. However, the most infallible quantitative data on phase equilibria are still obtained experimentally.

One of the simplest and most widespread experimental methods for constructing phase diagrams is a thermal analysis. A thermal curve or a thermogram is a time–temperature curve plotted at a constant cooling rate. To find the phase equilibrium point in the diagram, a change in the run of the curve should be recorded as the alloy passes from one region of the phase diagram to another. Figure 2.24 illustrates the thermal cooling curves for several alloys. As can be seen, the process of constructing the phase diagram is simple enough. The horizontal parts of the thermal curves correspond to crystallization of the eutectic and pure components.

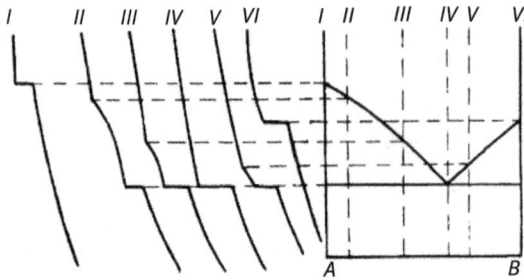

Figure 2.24: Thermal curves of alloys: constructing a eutectic phase diagram.

There are two reasons for the thermal analysis method to be not sufficient for constructing an accurate phase diagram. In the event of slightly differing the thermal properties of the phases, the sensitivity of the method is found to be weak. In this case, bends (kinks) of the thermal curves may be absent. Due to a slow cooling rate, the equilibrium cannot be achieved. As a consequence of the supercooling, the values of transformation temperatures become understated. Therefore, other more sensitive methods are applied for studying phase diagrams: microstructural and physical methods (dilatometric, resistometric, and magnetic ones). The X-ray structural method gains an advantage over the others because it allows one to ascertain, apart from transformation temperatures, the crystal structure of the transforming phases. To achieve an equilibrium state, all the methods require a sufficiently long holding period near the transformation temperatures. Therefore, in practice, phase diagrams have the abscissa axis only at elevated temperatures, and the low-temperature part of the diagram remains not constructed. This is due to low atomic mobility, and the equilibrium is difficult to achieve.

For building a phase diagram, the purity of the original components should be taken into account for obtaining accurate data. Even small concentrations of impurities can significantly affect the position of the diagram lines.

2.6 A diagram with limited solubility in the solid state

Let us look into a eutectic diagram formed by solid solutions based on pure components (Figure 2.25). The shape of the curve of the free energy of the limited solubility solution is shown in Figure 2.3.

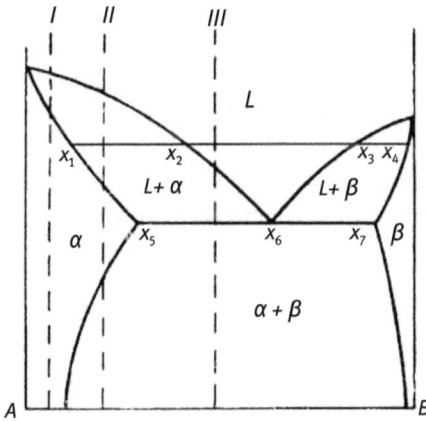

Figure 2.25: A limited solubility diagram for the solid state.

Figures 2.26 and 2.27 sketch the mutual arrangement of the free energy curves for a liquid solution and α and β solutions based on corresponding components A and B at temperatures above and equal to a eutectic temperature, respectively. In the latter case, the tangent line shares all the three free energy curves, that is, the liquid and solid solutions α and β are in equilibrium.

Consider the transformations in characteristic alloys of different compositions during cooling. In alloy I, crystallization occurs in a similar way as in an alloy with unlimited solubility components. In alloy II, after crystallization, another transformation is observed: an excess of component B is precipitated from the solid solution α in the form of particles of the solid solution β. This can be seen in the diagram when intersecting the solubility line of the component B in component A. The precipitation process or decomposition of a supersaturated solid solution is discussed in Chapter 6. In alloy III, after crystallizing the eutectics, the α and β solid solutions, when cooled, exhibit changes due to varying solubilities of the components.

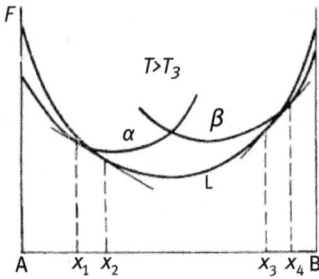

Figure 2.26: A limited solubility at the temperature above the eutectic temperature: free energy curves.

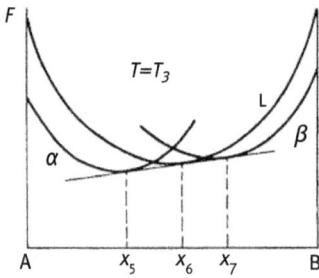

Figure 2.27: A limited solubility at the eutectic temperature: free energy curves.

In practice, diagrams with limited solubility of components in the solid state can often be faced. Examples of such diagrams are numerous; among them are Ag–Cu, Cr–Ni, Pb–Sn, Cd–Zn, and Cd–Pb.

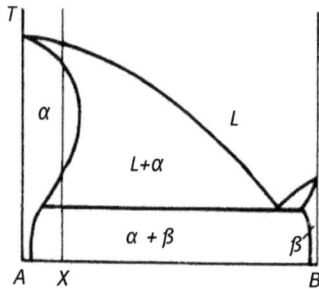

Figure 2.28: A retrograde solidus phase diagram.

A special case of a limited solubility diagram refers to a diagram with a retrograde solidus curve (Figure 2.28). The alloy of composition x first crystallizes to form the solid solution α. However, the shape of the solidus curve describes the crystallized solid solution as turning partly into a liquid again during cooling. This is retrograde melting. Upon further cooling, new crystals of the solid solution α are precipitated from the liquid. The latter crystallizes to form a mechanical mixture ($\alpha + \beta$) at the eutectic temperature. The phase region ($L + \alpha$) is referred to as the region of retrograde melting for alloys in a certain concentration range. Retrograde melting is characteristic of solid solutions based on a refractory component. The diagram of Ag–Pb serves an example of a diagram with a retrograde solidus curve.

2.7 A peritectic diagram

Peritectic diagrams are a particular case of diagrams with limited solubility of components in the solid state. Such diagrams are constructed for components differing significantly in melting points. Figure 2.29 presents a phase diagram with a peritectic reaction. Figure 2.30 shows the mutual arrangement of the free energy curves for a liquid solution and solid solutions α and β at a peritectic temperature. At the peritectic temperature, the tangent line to all the three free energy curves is common. However, unlike the eutectic diagram, the free energy curve for the liquid is located on one side from the free energy curves for solid solutions. This circumstance indicates the features of crystallization of these alloys.

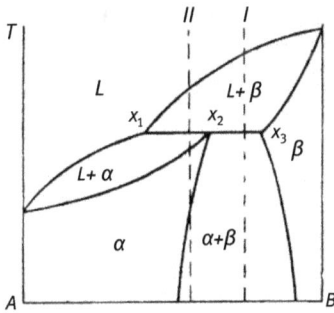

Figure 2.29: A peritectic diagram.

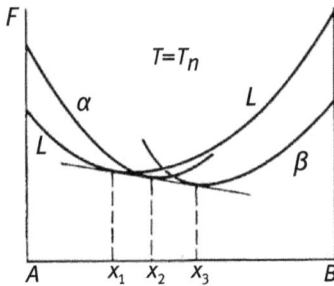

Figure 2.30: Free energy curves for liquid and solid solutions at the peritectic temperature.

Let us dwell on the transformations in alloys of different composition during cooling. In alloy I, when cooled, crystals of solid solution β are first precipitated. Upon reaching the horizontal (peritectic) line, the alloy consists of a liquid of composition x_1 and crystals of the solid solution β of composition x_3. It is worth noting that at this temperature there are three phases in equilibrium: a solid solution α of composition x_2 and the two phases mentioned above. For the alloy passing from the two-phase state $(L + \beta)$ into the three-phase $(L + \beta + \alpha)$ state, the number of degrees of freedom is zero. Consequently, the transformation occurs at a constant temperature. This transformation is the peritectic reaction. In other words, new α crystals are formed by interacting previously precipitated β crystals with a liquid. The peritectic reaction in this alloy proceeds until the complete disappearance of the liquid. It can be brought about in two ways or structural mechanisms. The first way includes the simultaneous growth of crystals of the solid solution α in a liquid and dissolution of β crystals. In addition, it is also possible for α crystals to grow on the surface of β crystals with simultaneous absorption of the liquid and the β crystals. Once finished, the peritectic reaction forms a mixture of the solid solutions $(\alpha + \beta)$. During further cooling, their chemical composition changes along the lines of limited solubility.

In alloy II, the peritectic reaction takes place until the complete disappearance of the β crystals. When cooled, the remaining part of the liquid crystallizes to form α crystals. The alloy comprises only crystals of solid solution α. Upon further cooling, the β crystals are precipitated from the α crystals; this is similar to that as in alloy II in Section 2.6. This means that the process of decomposition of the supersaturated solid solution α occurs. Peritectic diagrams are quite common. For example, these are Pt–Re, Cd–Hg, Co–Fe, and Fe–Ni diagrams.

2.8 Diagrams with intermediate phase

In binary systems, there often appear phases whose crystal structure differs from the structure of the components. These can be chemical compounds of strictly stoichiometric composition or, for example, intermetallic compounds existing in a certain concentration range. Such phases are called intermediate. The simplest diagram with a chemical compound is shown in Figure 2.31.

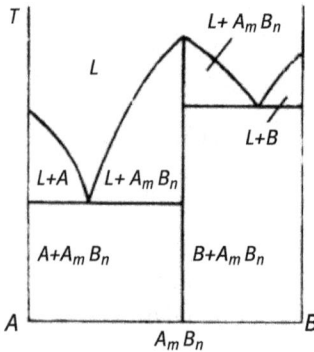

Figure 2.31: A diagram with a stable chemical compound of stoichiometric composition.

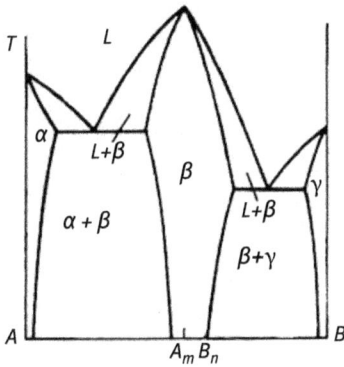

Figure 2.32: A diagram with a stable intermediate phase within a concentration range.

Each of the A and B components forms with the compound A_mB_n a diagram of the eutectic type. Thus, this diagram is divided into two parts. In each part, the A_mB_n compound plays the role of a component. If the region of the intermetallic compound extends over a concentration range or, otherwise speaking, the compound behaves like a limited solubility component, the diagram appears as in Figure 2.32. The above phase diagrams demonstrate the intermediate stable phases: during cooling, they crystallize from the liquid and undergo melting, when heated, without dissociating. Such intermediate phases are referred to as congruently melting phases. In other words, the chemical composition of the phase and liquid is the same.

If the intermediate phases are unstable, they dissociate when heated. The intermediate phase and liquid differ in chemical composition. In this case, the phase is called an incongruently melting phase. When cooled, the intermediate phase emerges in accordance with the peritectic reaction. Figure 2.33 depicts the phase diagram. When lowering the temperature, the component B is first precipitated from the alloy of composition I. Then the peritectic reaction results in the formation of the A_mB_n compound. Once the peritectic reaction is finished, the crystals of the compound A_mB_n and the crystals of the component B constitute the alloy. When cooling the alloy of composition II, the peritectic reaction leads to the precipitation of crystals of A_mB_n and the release of some liquid. Further, if the temperature continues to fall, the crystals of A_mB_n are precipitated from the liquid and then the remaining liquid crystallizes to form a eutectic $(A + A_mB_n)$. Thus, the alloy of composition II, when cooled, suffers both peritectic and eutectic transformations. In the event of replacing the intermediate phase by an intermetallic compound in the concentration range, the diagram has the form shown in Figure 2.34.

In binary systems, several intermediate phases can exist. Their phase diagrams are very complicated, and they should be divided into the constituent parts. For example, the Cu–Al, Fe–Al, and Co–Zn systems have five or more intermediate phases by both eutectic and peritectic reactions.

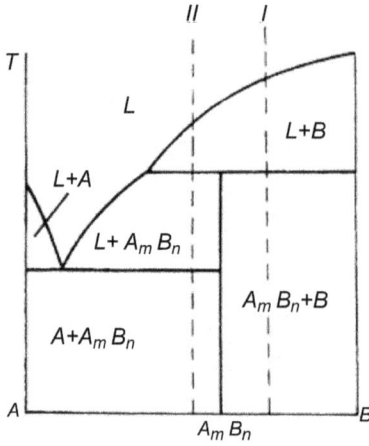

Figure 2.33: A diagram with an unstable chemical compound of stoichiometric composition.

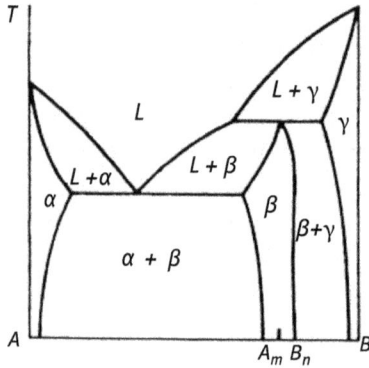

Figure 2.34: A limited solubility diagram with an unstable intermediate phase within a concentration range.

2.9 Diagrams with transformations at solid state

For many alloy systems, transformations in the solid state make phase diagrams complicated. As a rule, this is due to the polymorphism of components, for example, iron or titanium. The type of the phase diagrams depends on the number of polymorphic modifications of the components, on the influence of the chemical composition on the temperatures of polymorphic transformations, and on the magnitude of the solubility of one component in the polymorphic modifications of the others. There are various types of solid state transformation diagrams. They all consist of some number of characteristic segments with typical phase equilibria. Consider a few simple diagrams with component polymorphism.

Figure 2.35 involves a diagram with unlimited solubility of components in the liquid and solid states and with a polymorphic transformation of component B. As the concentration of component A increases, the temperature of the polymorphic transformation of

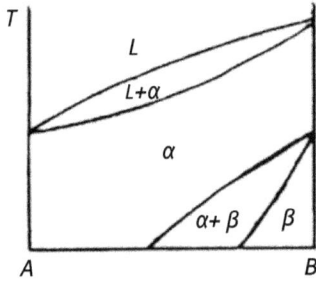

Figure 2.35: An unlimited solubility diagram with a polymorphous transformation in component B.

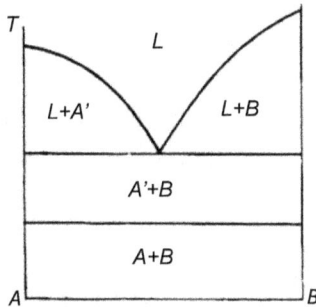

Figure 2.36: A eutectic diagram with a polymorphous transformation in component A.

the component B decreases. Thus, a two-phase region of solid solutions ($\alpha + \beta$) appears on the diagram. For example, Co–Ni and Ti–V systems have such a phase diagram.

Figure 2.36 shows a diagram with a eutectic and a polymorphic transformation of the component A. There is almost no solubility of components in the solid state. Therefore, a horizontal line in the diagram at the temperature of polymorphic transformation of the component A can be seen. This type of phase diagram is observed in Th–Ti and Pu–V systems. Since plutonium has several polymorphic modifications, the Pu–V system has several horizontal lines at temperatures of polymorphic transformations of plutonium.

Consider complicated diagrams when both components have polymorphism. Let high-temperature modifications of the components form a continuous series of solid solutions and low-temperature ones be limitedly soluble.

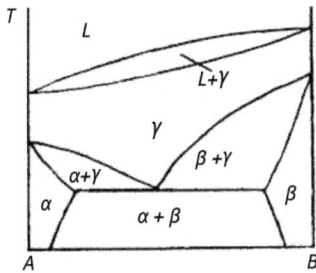

Figure 2.37: A eutectoid transformation diagram.

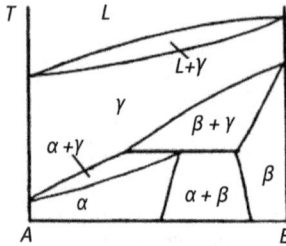

Figure 2.38: A peritectoid transformation diagram.

Figure 2.37 presents a limited solubility diagram for low-temperature modifications of the components. The α- and β-solid solutions based on these modifications form a part of the diagram similar to a eutectic. A γ-solid solution based on high-temperature modifications of the components is converted into a mechanical mixture of the solid solutions (α + β). Such a transformation is called eutectoid. In practice, a eutectoid transformation in the Fe–Fe₃C system is extremely important (see Section 7.4). Eutectoid transformations are also found in Cr–Ti and Ti–U systems. There are other possible transformations of low-temperature modifications of limitedly soluble components in the solid state. Figure 2.38 depicts a diagram, where the transformations in the solid state are similar to a peritectic. The already known γ-solid solution based on high-temperature modifications of the components reacts with the previously precipitated crystals of the solid solution β based on the component B. The reaction results in the formation of the α-solid solution based on component A. This transformation is called peritectoid.

A more complex diagram with polymorphic transformations of both components and the limited solubility of all the polymorphic modifications can be reviewed in Figure 2.39.

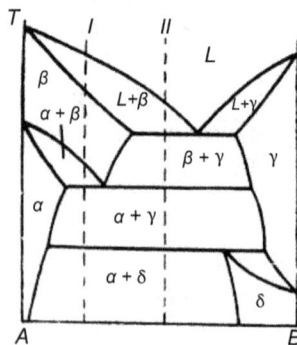

Figure 2.39: A phase diagram with polymorphous transformations in both components and their limited solubility.

Let us examine the sequence of the transformations during cooling in alloy I. Once crystallized within the interval (L + β), the alloy undergoes cooling in the β-region. Then, crystals of the α-solid solution are precipitated in the β-phase in the (α + β)

region. Afterward, a eutectoid transformation occurs, namely, the β-phase is transformed into a mechanical mixture of the α- and γ-phases. Upon further cooling, a peritectoid transformation takes place. In this case, the interaction of the α- and γ-phases
gives rise to a δ-phase formed.

Upon cooling, the transformations in alloy II are somewhat different. After precipitating the β-phase crystals from a liquid, eutectic crystallization happens followed by the formation of a mechanical mixture of the β- and γ-phases. After that,
the eutectoid transformation similar to as in alloy I is observed with converting the
β-phase into a mechanical mixture of the α- and γ-phases. Further, the δ-phase
emerges, caused by the peritectoid transformation, as in alloy I. As can be noticed,
in both alloys, the eutectoid and peritectoid transformations in the solid state arise
after crystallization. They are associated with the polymorphism of the components.
These transformations are accompanied by a significant grinding of the phase and
structural components, which is important for improving the mechanical properties
of industrial alloys. Heat treatment of steels and many non-ferrous alloys consists
in carrying out such transformations. The most important diagram in physical metallurgy is a Fe–Fe$_3$C phase diagram. It is the basis of transformations in steels and
cast irons. It will be discussed in Chapter 7.

2.10 Ternary phase diagrams

Multicomponent alloys are widely used in metallurgy, engineering, and other in-
dustries. The corresponding phase diagrams are necessary to know for analyzing
phase transformations during heat treatment of these alloys.

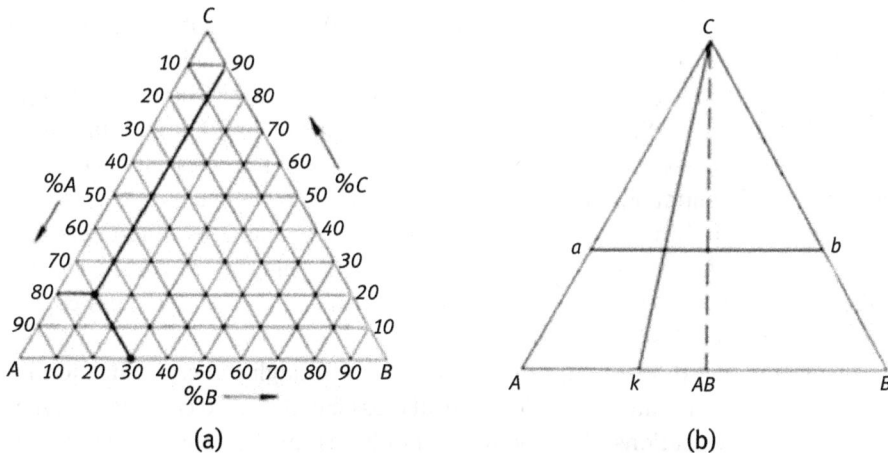

(a)   (b)

Figure 2.40: Depiction of alloy concentrations in a ternary system: (a) a Gibbs' concentration
triangle and (b) lines of characteristic vertical cross-cuts in the triangle.

Let us focus on the simplest ternary phase diagrams. In Figure 2.40a, a point on the equilateral triangle designates the chemical composition of a three-component alloy. The triangle bears the name of a concentration triangle or Gibbs' triangle. For example, an alloy containing 70% A, 10% B, and 20% C is shown. To illustrate the temperature change, it is required to draw an axis perpendicular to the concentration triangle. As a result, a three-component phase diagram is a spatial one inside a triangular prism whose side faces are binary diagrams.

Figure 2.41 shows a diagram with an unlimited solubility of the components in the liquid and solid states.

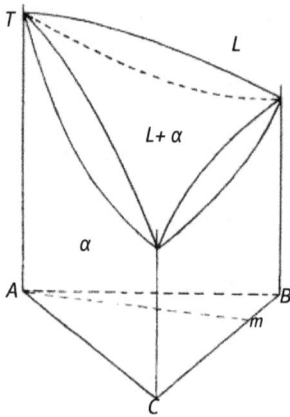

Figure 2.41: A spatial unlimited solubility diagram.

The melting points of the components reside on the edges of the prism. The prism side faces are of binary unlimited solubility diagrams ("cigars"). The liquidus and solidus surfaces in the ternary phase diagram are bounded by the liquidus and solidus lines of the binary diagrams. The polythermal (vertical) cross-cuts of the diagram are convenient to apply for a pictorial demonstration of the transformation temperatures during cooling (heating) of specific alloys. Typical cross-cuts can be made through the top of the concentration triangle or parallel to its side (Figure 2.40b). If the cross-cut passes along the Ck line, the alloys lying in the plane of the cross-cut have a constant ratio of concentrations of the components A and B. A cross-cut passing through the concentration of the compound, for example, along the C–AB line refers to a quasi-binary one. In the plane of the cross-cut passing along the ab line, the content of the component C is constant. Figure 2.42 gives an example of two cross-cuts of such an unlimited solubility diagram.

The isothermal (horizontal) cross-cut is drawn through the crystallization region ($L + \alpha$). It is worth noting that the vertical cross-cut allows no conoids for determining the concentrations of the equilibrium phases. In the general case, this is because of location of no conoids in the cross-cut plane. The conoids can be run on

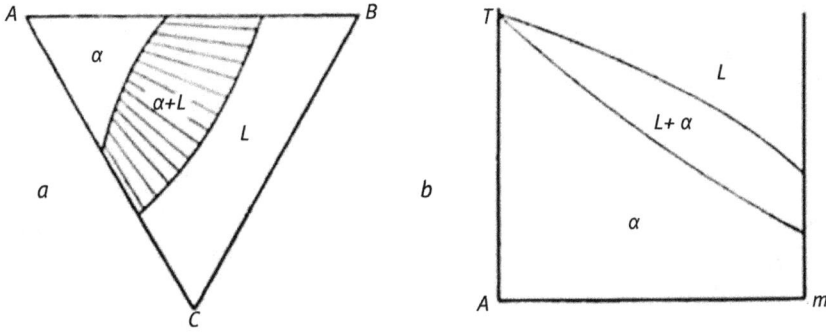

Figure 2.42: An unlimited solubility diagram: (a) A horizontal cross-cut drawn through the two-phase region ($L + \alpha$) and (b) a vertical cross-cut drawn along the line Am.

the horizontal cross-cut. In doing so, however, the concentration of the components in one of the phases should be assigned.

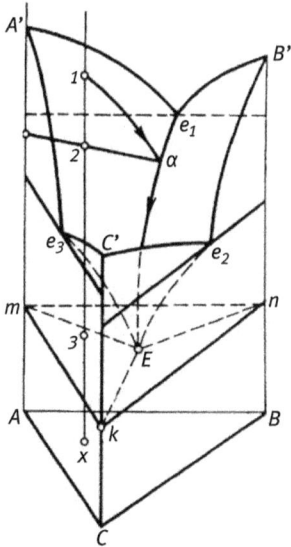

Figure 2.43: A spatial ternary eutectic phase diagram.

Figure 2.43 presents a ternary eutectic phase diagram. Binary diagrams of the eutectic type are depicted on the side faces of the prism. In two-component systems, two liquidus lines of each component, for example, the $A'e_1$ and A' lines, can be passed through the liquidus surface of this component in a ternary system.

In the ternary system, the total liquidus surface consists of three surfaces of the primary crystallization of each component. Every two surfaces of the liquidus of primary crystallization intersect along the lines of binary eutectic e_1E, $e2E$, and $e3E$, emanating from the eutectic points of the two-component systems. The three surfaces of the primary crystallization of the components A, B, and C, as well as the three lines of binary eutectics, intersect at the point of the ternary eutectic E.

Let us explore the process of crystallization of an alloy composition of the point x. When the liquidus surface crosses the point 1, the component A begins precipitating; the chemical composition of the liquid changes along the line $1a$. At point 2, the binary eutectic $(A + B)$ of the composition of the point a crystallizes; the composition of the liquid changes along the e_1E line. When crystallizing the binary eutectic, three phases are balanced $(A, B, \text{and liquid } L)$. Therefore, according to the phase rule, in the ternary system, the binary eutectic crystallizes as the temperature varies. With a further decrease in temperature at point 3, the ternary eutectic $(A + B + C)$ of the composition of the point E crystallizes. The process proceeds at a constant temperature because four phases $(A + B + C + L)$ are in equilibrium. In any three-component alloy, crystallization lasts until the formation of a ternary eutectic. In the phase diagram, the horizontal plane mnk of the ternary eutectic crystallization passes through the point E. All ternary alloys are in the solid state below this plane.

Let us examine the projection of the spatial regions of the ternary diagram on the plane of the concentration triangle (Figure 2.44a).

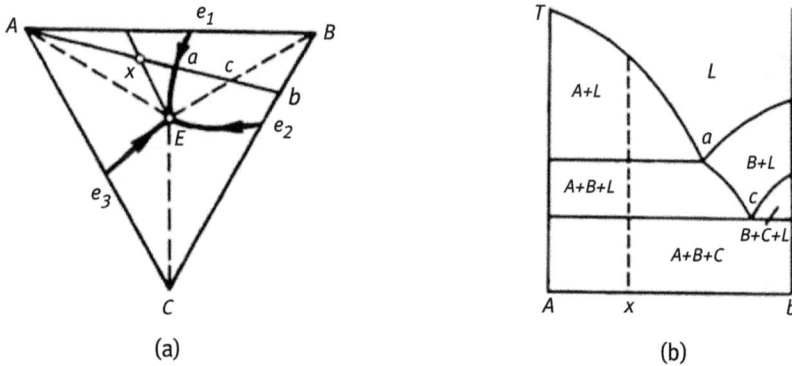

Figure 2.44: Cross-cuts in a ternary eutectic diagram; (a) a horizontal cross-cut and (b) a vertical cross-cut.

The e_1E, e_2E, and e_3E lines are the projections of binary eutectic lines. The regions Ae_1Ee_3, Be_1Ee_2, and Ce_2Ee_3 present the regions of primary crystallization of the components A, B, and C, respectively. The triangular regions ABE, BCE, and ACE involve the regions of crystallization of binary eutectics $(A + B)$, $(B + C)$, and $(A + C)$, respectively. The above-described crystallization process for the alloy of the composition of the point x is clearly outlined on the plane of the concentration triangle. At the stage of the primary crystallization, the component A is isolated; the composition of the liquid changes along the line Aa. Once the concentration of the liquid reaches the point a, the $(A + B)$ binary eutectic begins to form; the liquid composition changes along the e_1E line. The crystallization ends with the formation of a ternary eutectic. The successive passage of the phase regions and the change in concentration can be seen on the concentration triangle plane. However, the transformation temperatures are impossible to depict. They should be shown through constructing vertical cross-cuts. Figure 2.44b sketches such a vertical cross-cut marked with the Ab line on the concentration triangle plane. The position of the alloy composition of the point x is fixed on the cross-cut. Thus, the ternary diagram consists of eight spatial phase regions: the liquid solid solution region, the three primary crystallization regions of each component, three binary eutectic crystallization regions, and the ternary eutectic crystallization region.

The phase diagram becomes more complicated if the A, B, and C limited soluble components form the α, β, and γ solid solutions, and the ternary eutectic is produced from solid solutions (Figure 2.45).

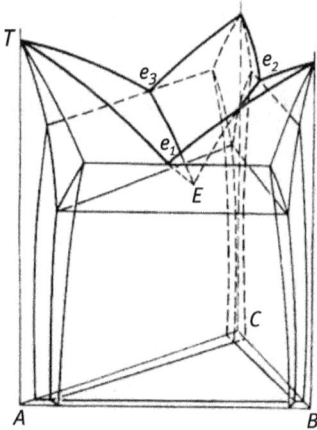

Figure 2.45: A spatial phase diagram with a limited solubility of components and with a ternary eutectic.

As in the previous case, the point E indicates the ternary eutectic position; the lines of binary eutectics are e_1E, e_2E, and e_3E. Solid solutions form three spatial phase regions that are absent in the previous spatial diagram. Other three spatial phase regions create binary eutectics consisting of solid solutions: $(\alpha + \beta)$, $(\beta + \gamma)$, and $(\alpha + \gamma)$. Thus, this ternary diagram includes 14 spatial phase regions. The horizontal cross-cuts of the diagram (Figure 2.46) present the arrangement of these regions. One cross-cut is made below the ternary eutectic temperature (Figure 2.46a), and another is drawn above this temperature, but below the temperature of the low-melting binary eutectic (Figure 2.46b).

To quantitatively determine the ratio of the phases in the two-phase region of the ternary diagram, the lever rule is applied, similar to as in a binary system. If the phases of composition d and e are in equilibrium on a horizontal cross-cut in a two-

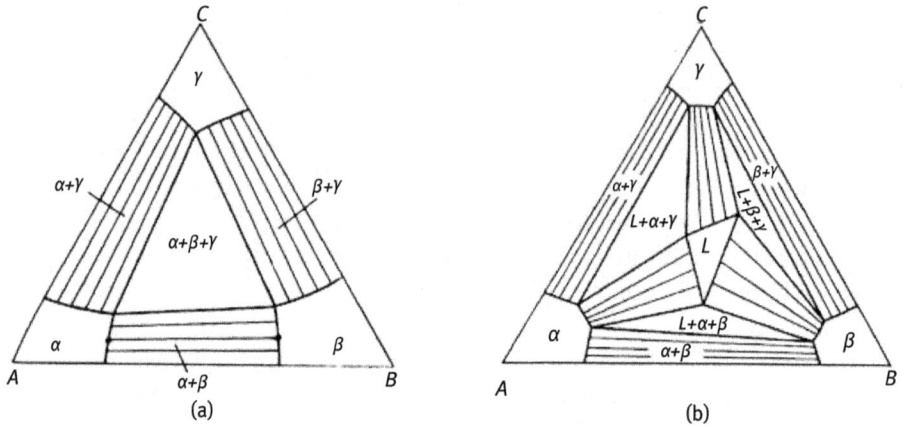

Figure 2.46: Horizontal cross-cuts in a limited solubility ternary diagram: (a) below the ternary eutectic temperature and (b) between the eutectic temperature and the lowest melting binary-eutectic temperature.

phase region in an alloy of composition r, the amount of the phase d is calculated via the re/de ratio (Figure 2.47a).

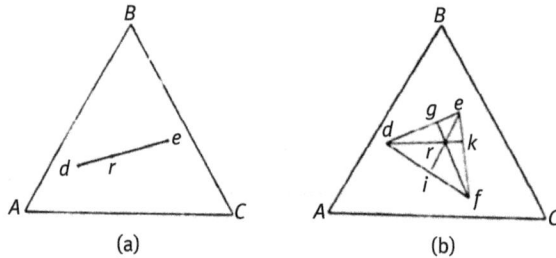

Figure 2.47: For the lever rule (a) and for the center of gravity rule (b).

If the alloy of composition r resides in a three-phase region, the quantitative ratio of the phases is determined by the center of gravity rule. Thus, the phases of composition d, e, and f are in equilibrium, and the amount of the phase d is determined by the rk/dk ratio (Figure 2.47b).

For validation of building phase diagrams, the law of adjacent state spaces is reasonable to use. It was mentioned in Section 2.4. The law reads that two state spaces can have a common border or be adjacent over a certain separation surface only in the event of differing the number of phases in these spaces more than a unit. A single-phase state, for example, can border over the separation surface only with a two-phase one. A two-phase state can be adjacent only with a single-phase or three-phase one. The rule may seem to be contradictious to the case when the crystallization space of a $(A + B + L)$ binary eutectic and a $(A + B + C)$ three-phase space, corresponding to complete solidification, contact between each other in the ternary system (Figure 2.44b). However, it should be kept in mind that these spaces are separated by the $(L + A + B + C)$ ternary eutectic crystallization plane that gives no spatial representation.

The top of the section dealt with the use of multicomponent alloys in engineering. When described, the phase diagrams of such alloys face geometric difficulties. To illustrate a phase diagram for a four-component system, a three-dimensional measurement space is not sufficient. In the three-dimensional space, only the chemical composition of the alloys can be represented if one images the system of the alloys as a tetrahedron of concentrations. The vertices of such a tetrahedron are the components the alloys are made of. The edges of the tetrahedron include six binary systems; its side faces can place four ternary systems. To reveal a temperature change in a four-component system, a fourth dimension is required. Therefore, when constructing the diagrams, convenient cross-cuts are forced to use for labeling the concentration of one or more components. Furthermore, the diagrams should be simplified depending on the chemical composition of a particular alloy and its heat treatment conditions. Detailed information on the structure of various binary, ternary, and multicomponent diagrams can be found in the special literature.

2.11 *T–P* phase diagrams

Some high-pressure industrial technologies affect the phase equilibria, and the influence of the pressure becomes decisive. Due to high pressure, the position of the lines of equilibrium diagrams changes. This causes new high-pressure phases to form, which is absent at normal pressure. Pressure has a strong influence on the establishment of equilibrium and affects the kinetics of phase transformations. According to the Clapeyron–Clausius equation, the stronger the effect of pressure, the greater the volume effect of phase transformations is. As a consequence, the equilibrium diagrams, when constructed, need to take pressure changes into account, along with temperature and concentration. Obviously, in this case, a two-component system should be depicted in the three-dimensional measurement space. For a one-component system, *T–P* phase diagrams are displayed on a plane. The present section covers the simplest examples of *T–P* diagrams in one-component systems.

In thermodynamics textbooks, the description of *T–P* diagrams usually begins with a phase diagram for water. Three lines of the diagram separate the phase regions of the three aggregative states of water: liquid, solid, and gaseous ones (see the scheme in Figure 2.48).

At the triple point (at 0 °C and a pressure of 1 atm), all the lines intersect, and this corresponds to the invariant equilibrium of the three phases. The water–vapor equilibrium line has another characteristic (critical) point, where the difference between the gas (vapor) and the liquid disappears.

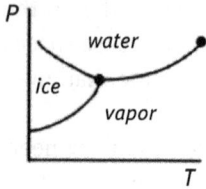

Figure 2.48: The *T–P* water phase diagram.

Constructing the *T–P* iron phase diagram is a result of many experimental investigations (Figure 2.49).

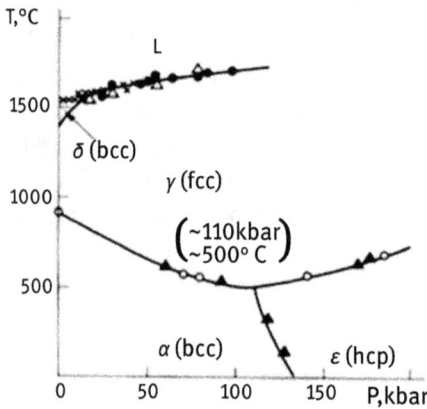

Figure 2.49: The *T–P* iron phase diagram.

Under pressure, the melting temperature increases. The triple point for transformations in the solid state (phases α, γ, and ε) is located at ~ 500 °C and a pressure of ~ 11 GPa (110 kbar). The high-pressure phase ε has a hexagonal structure whose specific volume is less than the specific volumes of the other iron phases (bcc α-phase and fcc γ-phase). The ε-phase formed under pressure disappears with decreasing pressure, and a hysteresis loop of the transformations α → ε and ε → α emerges. It should be noted that the ε-phase can be preserved after releasing the pressure (at least partially) if the pressure exposure is accompanied by shear deformation. The simultaneous impact of high pressure and plastic deformation on phase transformations refers to a special section of high-pressure physics.

Figure 2.50 shows a portion of the *T–P* titanium diagram for transformations in the solid state. The triple point (hcp α-phase, bcc β-phase, and ω) for titanium is located at ~ 600 °C and pressure of ~ 9 GPa (90 kbar). The ω high-pressure phase has a complicated crystal structure.

The *T–P* carbon diagram is of great scientific and practical interest. At certain values of pressures and temperatures, carbon exists in the form of diamond. The

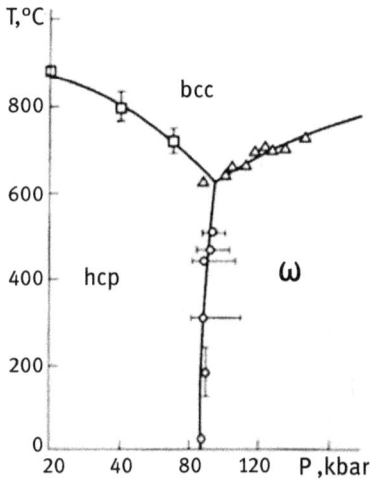

Figure 2.50: The *T–P* titanium phase diagram.

graphite–diamond transition under static conditions requires a high temperature of 2,800–3,300 K and a pressure of 9–10 GPa. In a dynamic mode, the transition occurs at 300 K and 22 GPa.

The problems of phase transformations in solids under high pressure are discussed in detail in a recently published book by Blanc and Estrin in 2011.

2.12 Metastable diagrams

As mentioned in Section 2.5, to achieve equilibrium at low temperatures, extremely long holding periods are required. So, the low-temperature part of the diagrams often remains unbuilt. For compensating this deficiency, metastable diagrams (realization diagrams) are utilized. They show the temperature–concentration regions of phase states under real heat treatment conditions. For example, Figure 2.51 presents the iron angle of the metastable phase Fe–Ni diagram for transformations in the solid state (Hansen and Anderko).

With continuous cooling, the diagram indicates the temperature lines near the beginning and near the end of the $\gamma \rightarrow \alpha$ transformation (10% and 90% of the transformation, respectively). With continuous heating, the lines near the temperatures of the beginning and end of the reverse $\alpha \rightarrow \gamma$ transformation can also be seen. The transformation temperatures almost do not change in the rate range from 2 to 150 deg/min. In the interval between the *a* and *c* curves, the alloys can be either in the α or γ state, depending on heat treatment. It can be noticed that the hysteresis of the $\gamma \rightarrow \alpha$ and $\alpha \rightarrow \gamma$ transformations increases with increasing nickel concentration and reaches 400 °C. Figure 2.52 demonstrates the iron angle of the metastable phase Fe–Mn diagram (Hansen and Anderko) for transformations in the solid state. As

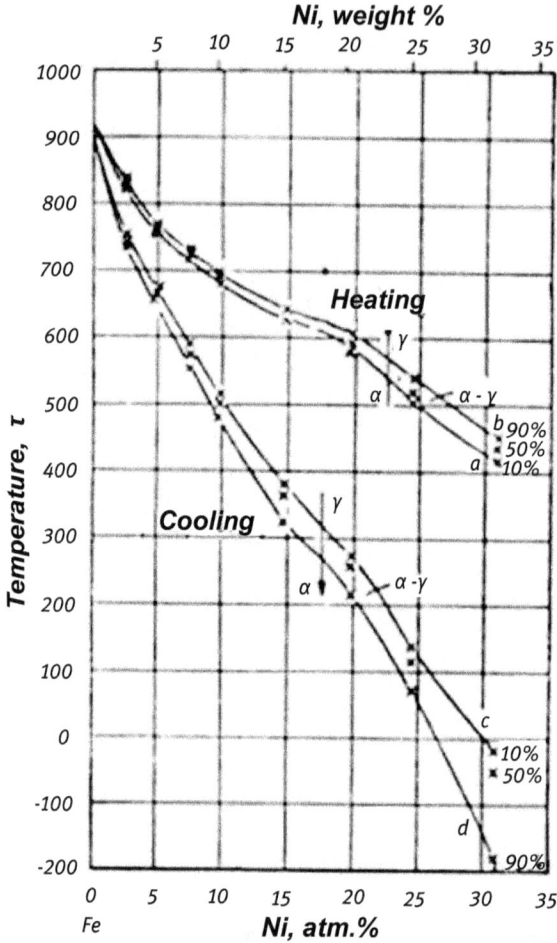

Figure 2.51: The iron angle in the metastable Fe–Ni diagram.

in the previous example, the temperature lines of the γ → α transformation during cooling and the temperature lines of the α → γ transformation during heating are shown. In the system Fe–Mn, at certain concentrations of manganese, a metastable hcp ε-phase is formed. The diagram specifies the γ → ε transformation lines during cooling and the reverse ε → γ transformation lines during heating. Metastable diagrams help choose heat treatment conditions for real alloys.

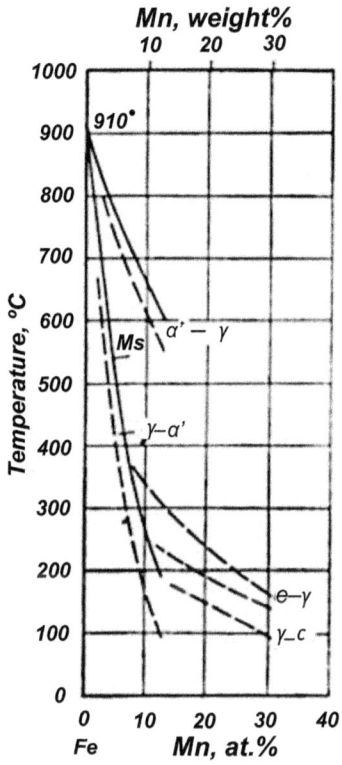

Figure 2.52: The iron angle in the metastable Fe–Mn diagram.

Chapter 3
Crystallization

3.1 The liquid and solid states of a metal

Crystallization (solidification) is a very widespread case of the phase transformation. There are many practical fields such as smelting and casting of steel, metals, and alloys, the foundry industry for manufacturing various cast products, and the growth of single crystals under slow cooling, where the phase transformations can be observed. The transition from the liquid to solid state occurs during cooling due to heat dissipation conditions that determine the structure metal billets obtained.

Most characteristics of the liquid state of a metal are closer to a solid one than to gaseous. Only van der Waals forces act between atoms of metallic vapor. In a liquid metal, as in a solid one, valence electrons are collectivized, and the main type of bond is metallic. The forces of interatomic bonding decrease weakly during melting and dramatically under evaporation. The latent heat of melting is not more than 10% of the evaporation heat. The main difference between the structure of a liquid metal and the structure of a metal in a solid state consists in the lack of a long-range order in the arrangement of atoms. In a liquid metal, the short-range order is preserved. The latter is characterized by the most probable shortest interatomic distance and the average number of the nearest neighbors of an arbitrary atom (the coordination number). The coordination numbers of liquid metals near the melting point are not very different from 12, especially for metals with a dense packing of atoms (Cu and Ni).

A real molten metal is not a homogeneous liquid. It contains dispersed particles of refractory substances (oxides, nitrides, carbide, etc.) formed during smelting. These particles have a strong influence on the crystallization process and the structure of the ingot (casting). A homogeneous free impurity metal can be produced by special methods of preparation of alloys such as smelting in an inert atmosphere and vacuum arc melting.

Thermodynamics reads that free energy (the Gibbs energy) decreases with increasing temperature. The Gibbs energy drops more intensively for the liquid state than for the solid state (see Chapter 2). The free energy dependencies on temperature for a liquid and a solid intersect so that a thermodynamically stable state is characteristic of a solid at low temperatures; however, a liquid is stable at high temperatures (Figure 3.1).

The intersection point corresponds to an equilibrium temperature of the two phases, T_0. In order for the crystallization to begin, it is necessary that the liquid (melt) should be overcooled below T_0. The difference $\Delta F = F_{liq} - F_{solid}$ is a thermodynamic stimulus (a driving force) for the crystallization. The greater the degree of overcooling, ΔT, the greater the driving force ΔF is.

https://doi.org/10.1515/9783110758023-003

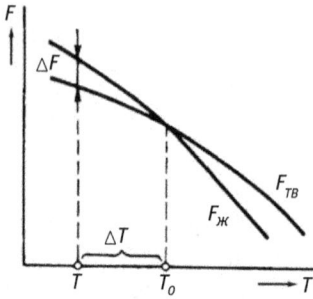

Figure 3.1: Free energies of a liquid and a solid, depending on the temperature.

The kinetics of crystallization is determined by the rate of nucleation of the crystallization centers and their growth rate. Figure 3.2 sketches the dependencies of the number of crystallization centers and the linear growth rate of crystals on the degree of supercooling (the Tamman curves).

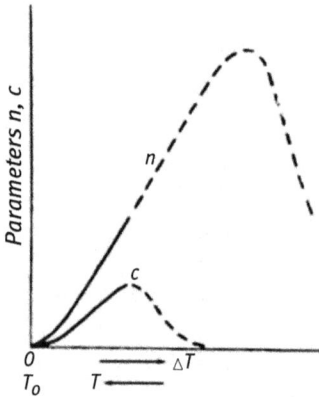

Figure 3.2: Dependencies of the crystallization centers (n) and the linear growth rate of crystals on supercooling.

With an increase in the degree of supercooling, both crystallization parameter values first rise and then diminish. The center number reaches its maximum value at higher supercooling rather than the growth rate. An explanation of the course of these dependencies is given below.

3.2 The theory of homogeneous nucleation

To begin with, we consider the theory of homogeneous nucleation. A crystallization center (nucleus) to emerge in the volume of a liquid metal requires an energy fluctuation. As a result of an increase in the free energy of the part of the liquid, the nucleus can turn into a crystal. Frenkel called such a fluctuation as heterophase. The probability of its occurrence is

$$W \sim A \exp \left(\frac{-\Delta\Phi}{kT} \right),$$

where $\Delta\Phi$ is the increase in free energy due to the appearance of this fluctuation, A is a constant, and k is the Boltzmann constant. When formed, the nucleus spends the free energy of a given volume. However, the surface energy increases due to the appearance of the crystal/liquid separation surface. The total change in the free energy to create the crystallization center due to heterophase fluctuations can be presented as

$$\Delta\Phi = - V \times \Delta f_V + S\sigma,$$

where V is the volume of the crystallization center, Δf_V is the difference between the free energies of the liquid and solid phases per unit volume, S is the surface of the nucleus, and σ is the surface tension:

$$\Delta f_V = \frac{\Delta F}{V} = \frac{F_{\text{liq}} - F_{\text{sol}}}{V},$$

where F_{liq} and F_{sol} are the free energies of a liquid and a solid in volume V.

For a spherical nucleus, the following formula holds true:

$$\Delta\Phi = - \frac{4}{3}\pi r^3 \Delta f_V + 4\pi r^2 \sigma,$$

where r is the radius of the nucleus.

At low degrees of supercooling, the second summand prevails over the first one. This means that the formation of the crystallization center is energetically unfavorable because of an increase in the free energy. As the temperature falls further, the value of Δf_V grows. Moreover, the first term contains the radius of the nucleus raised to the third power, and the second term contains the radius squared of the nucleus. Consequently, the first term begins to prevail over the second one. With equality of the first and second terms ($\Delta\Phi = 0$), the radius of the nucleus becomes critical ($r = r_{\text{crit}}$), and the nucleus tends to a spontaneous growth. In this case, the free energy decreases. If the nucleus size is less than the critical one, then the nucleus is thermodynamically unstable and disappears. A change in the free energy of the system for creating the crystallization nuclei of various sizes is shown in Figure 3.3.

As the temperature decreases, the size of the critical nucleus, r_{crit}, decreases. Let us find r_{crit} from the condition $(d\Delta\Phi/dr) = 0$:

$$- 4\pi\Delta f_V r^2 = 8\pi\sigma r = 0;$$

$$r_{\text{crit}} = \frac{2\sigma}{\Delta f_V}.$$

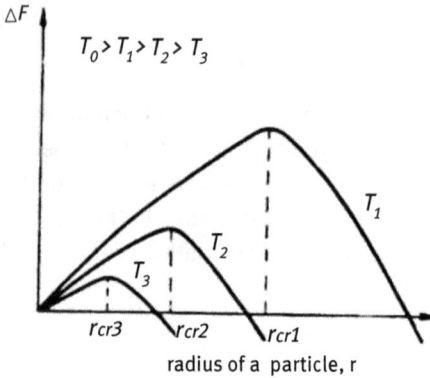

Figure 3.3: Change in free energy of a system upon forming crystallization nuclei of different sizes.

For a nucleus in the shape of a cube with a side length of a, we have

$$a_{\text{crit}} = \frac{4\sigma}{\Delta f_V}.$$

For the emergence of a spherical nucleus of the critical size r_{crit}, the increment $\Delta\Phi_{\text{crit}}$ appears as follows:

$$\Delta\Phi_{\text{crit}} = -\frac{4}{3}\pi r_{\text{crit}}^3\,\Delta f_V + 4\pi r_{\text{crit}}^2\sigma = -\frac{32}{3}\pi\frac{\sigma^3}{\Delta f_V^2} + 16\pi\frac{\sigma^2}{\Delta f_V^2} = -\frac{1}{3}16\pi\frac{\sigma^3}{\Delta f_V^2}.$$

In this case, the work of forming the surface of the spherical nucleus of the critical size, A_{crit}, is

$$A_{\text{crit}} = \sigma S_{\text{crit}} = 16\pi\frac{\sigma^3}{\Delta f_V^2}; \qquad \Delta\Phi_{\text{crit}} = \frac{1}{3}A_{\text{crit}} = \frac{1}{3}\sigma S_{\text{crit}}.$$

The formation of the critical size of nucleus turns out to be energetically unavailable. The deficiency equal to 1/3 of the work of forming the surface of the nucleus is replenished by fluctuation processes.

Consider how temperature affects the number of the crystallization centers. It would seem that the number of centers will continuously increase with an increase in the degree of supercooling, $\Delta T = T - T_0$. However, the atomic mobility decreases as the temperature drops. In other words, the self-diffusion coefficient also lowers. As a consequence, the fluctuations prevent the crystallization centers to appear. The presence of two mutually opposite tendencies leads to a curve with a maximum in the dependence of the number of crystallization centers (n) on supercooling (ΔT).

The number of the nucleus centers that arise in a unit volume per unit of time is expressed by the following equation:

$$n = K \exp\left(-\frac{\Delta\Phi_{crit}}{kT}\right) \exp\left(-\frac{U}{kT}\right) = K \exp\left(-\frac{\Delta\Phi_{crit} + U}{kT}\right),$$

where K is the proportionality coefficient and U is the activation energy required for an atom to move from one phase to another. The first factor takes into account the change in energy fluctuations needed for the formation of nuclei of a critical size. The second factor makes an allowance for the change in atomic mobility.

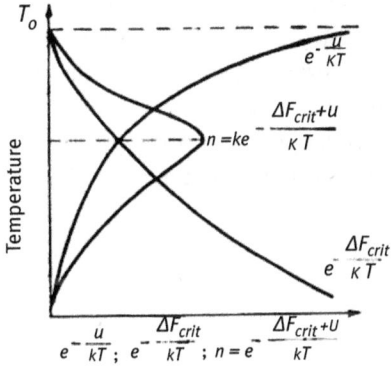

Figure 3.4: Supercooling temperature dependence of the number of crystallization nuclei.

Figure 3.4 outlines the dependencies of these factors and the general dependence of the number of nuclei on the supercooling temperature.

3.3 Heterogeneous nucleation

The previous section covered the formation of crystallization centers in a homogeneous melt. Now we look into heterogeneous nucleation. The process of nucleation on a certain inclusion, for example, on a solid particle, can significantly reduce the change in free energy due to a decrease in surface energy. When nucleating on the finished surface, the change in the free energy is equal to

$$\Delta\Phi_{fin.surf.} = \frac{1}{3}(\sigma_1 S_1 + \sigma_2 S_2),$$

where S_1 is the surface area between the nucleus and the liquid phase, S_2 is the surface area between the nucleus and the inclusion, σ_1 is the surface tension at the separation boundary between the nucleus and the liquid phase, and σ_2 is the surface tension at the separation boundary between the nucleus and the inclusion. To facilitate the nuclei formation on the finished separation surfaces, the condition $\Delta\Phi_{fin.surf.} < \Delta\Phi_{hom}$ should be satisfied. Otherwise speaking, $\sigma_1 > \sigma_2$. The smaller the surface tension at the nucleus/inclusion boundary as compared with that at the nucleus/liquid phase boundary, the more active role the nucleation on the finished surface would play.

The surface energy of the interphase nucleus/inclusion boundary correlates to the following statement. The surface energy is smaller, the greater the similarity in the arrangement of atoms on the contacting surfaces and the similarity in the type of interatomic bond on both sides of the separation boundary. The minimum surface tension corresponds to the similarity of crystal lattices on contiguous planes. The Dankov–Konobeevsky principle claims: "The transformation on the surface of a solid body proceeds in such a way that the configuration (arrangement) of the atoms of the original solid phase should be preserved (or almost preserved) in the new solid phase. The new phase lattice arisen during this crystallization process meets the crystal lattice of the old phase the same as similar crystal planes whose parameters differ minimally from each other." The Dankov–Konobeevsky principle or the principle of orientational and dimensional correspondence holds true for both a crystallization process and other phase transformations.

3.4 Crystallization kinetics

3.4.1 Growth kinetics of nucleus centers

A crystallization nucleus grows due to the attachment of individual atoms or their aggregates to its surface. With sufficiently slow cooling, the growth happens layer by layer. A two-dimensional nucleus emerges on the surface of the core–nucleus, capable of growing spontaneously. The role of the two-dimensional nucleus, as well as of a three-dimensional nucleus in the process of homogeneous nucleation, is the same. The change in free energy as the two-dimensional nucleus emerges is

$$\Delta\Phi = -\Delta V \times \Delta f + \Delta S \times \sigma,$$

where ΔV is the volume of a two-dimensional nucleus and ΔS is an increase in its surface.

Suppose the nucleus has the shape of a thin parallelepiped with a thickness δ and a cross section in the shape of a square with a side length of a, then:

$$\Delta\Phi = -a^2\delta \times \Delta f + 4a\delta\sigma.$$

Figure 3.5 illustrates the dependence of the free energy change $\Delta\Phi$ on the size a of the two-dimensional nucleus.

The plot has a maximum at the critical size of the two-dimensional nucleus, a_{crit}. This size can be found in an analogous way as in the case of the critical size of a three-dimensional nucleus:

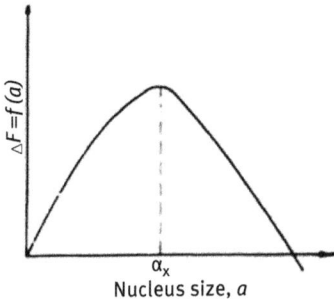

Figure 3.5: The dependence of the free energy change $\Delta\Phi$ on the size of a two-dimensional nucleus.

$$\frac{d\Delta\Phi}{da} = -2a\delta \times \Delta f + 4\delta\sigma = 0; \quad a_{crit} = \frac{2\sigma}{\Delta f}.$$

As a two-dimensional nucleus of a critical size a_{crit} emerges, the increment $\Delta\Phi_{crit}$ is

$$\Delta\Phi_{crit} = -\frac{4\sigma^2}{(\Delta f)^2}\delta \times \Delta f + 4\frac{2\sigma}{\Delta f}\delta\sigma = -\frac{4\sigma^2\delta}{\Delta f} + \frac{8\sigma^2\delta}{\Delta f} = \frac{4\sigma^2\delta}{\Delta f} = \frac{1}{2}\sigma \times \Delta S; \quad \Delta\Phi_{crit} = \frac{1}{2}A_{crit}.$$

It is seen that the formation of the two-dimensional nucleus of a critical size turns out to be energetically unavailable. The deficiency equal to 1/3 of the work of forming the nucleus surface is replenished by fluctuation processes similar to the case of a three-dimensional nucleus.

The number of two-dimensional nuclei, as well as the number of three-dimensional nuclei, has the dependence on the degree of supercooling in the form of a curve with a maximum. Therefore, the temperature curve of the linear growth rate has the shape as in Figure 3.6.

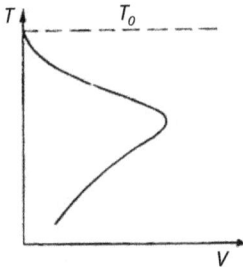

Figure 3.6: Temperature dependence of the linear growth rate of two-dimensional nuclei.

Frenkel showed that, against the maximum number of three-dimensional nuclei, the maximum number of two-dimensional nuclei can be created at a smaller degree of supercooling. The reason for two-dimensional nuclei to grow is to require smaller energy fluctuations than that for three-dimensional ones. So, the dependencies of the number of crystallization centers (n) and the linear crystal growth rate (c) on the degree of supercooling have maxima at different supercoolings (Figure 3.2).

Once the two-dimensional nucleus emerges, a new layer grows rather easily. Figure 3.7 displays a diagram illustrating the different positions of two-dimensional nuclei on the face of a growing crystal.

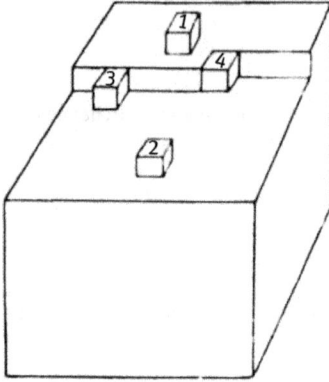

Figure 3.7: Different positions of two-dimensional nuclei on the face of a growing crystal.

Unlike the emergence of the nuclei 1 and 2, the formation of nucleus 3 is facilitated since it requires a smaller area of the new surface. Consequently, nucleus 3 has a smaller critical size. The birth of nucleus 4 needs no increase in the surface area; it occurs barrier-freely, without energy fluctuations. It follows from the diagram that the emergence of nuclei of types 1 and 2 provokes the growth of each new layer. This statement holds true for small degrees of supercooling. At large supercooling, the nuclei of types 1 and 2 appear even before the layer is filled. In this case, the growth mechanism changes from a layer-by-layer growth to the growth along the normal to the interphase surface. It is worth emphasizing that two-dimensional nuclei are most easily formed on planes with dense packing of atoms.

3.4.2 Growth mechanisms

Let us present one of the possible classifications of growth mechanisms.

1. *Growth in the formation of two-dimensional nuclei:* layer-by-layer growth (synonyms: stepped or lateral ones). Figure 3.7 illustrates the growth of nuclei 3 and 4.
   Growth rate V according to Volmer is

$$V \sim \exp\left(-aT/b\Delta T\right),$$

where a and b are constants, T is the crystallization temperature, and ΔT is the supercooling value.

2. *Growth through screw dislocations.* In the output of the dislocation to the surface of a growing crystal, a dislocation creates a step. One revolution around the output

point moves the growing surface by one interatomic layer. Growth rate according to Turnbull is

$$V \sim \Delta T^2.$$

In fact, this growth mechanism is a variation of the previous one.

3. *Continuous growth.* Contrary to lateral growth, it is a growth normal to the interphase separation surface. Figure 3.7 displays the growth of the nuclei 1 and 2. Growth rate according to Turnbull is

$$V \sim \Delta T.$$

Figure 3.8 shows the growth rate dependencies on supercooling for the growth mechanisms mentioned above.

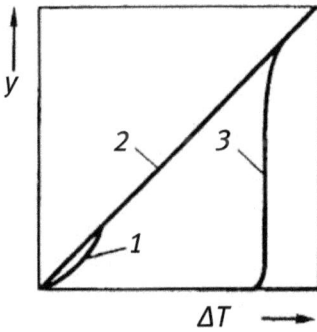

Figure 3.8: Growth rate dependencies on supercooling for various growth mechanisms: 1, screw dislocation growth; 2, continuous growth; 3, two-dimensional nucleation.

4. *Dendritic growth.* It is a special form of growth, a kind of continuous one. Figure 3.9 gives an example of dendritic growth: the successive stages for a dendritic crystal of organic matter to grow are seen.

3.4.3 Formal kinetics

The kinetics of crystallization depends on the number of arising three-dimensional nuclei (n) and on their linear growth rate (c) or on the number of two-dimensional nuclei:

$$V_t = V_0 \left(1 - \exp\left(-\frac{\pi}{3} c^3 n t^4 \right) \right),$$

where V_t is the amount of the crystallized substance during time t. This is an approximate calculation of Kolmogorov for spherical nuclei. Attention should be drawn to the fact that the expression for crystallization kinetics contains the growth rate raised in the third power, and the nucleation rate is in the first power. Consequently, a change in the growth rate has a stronger influence on the crystallization

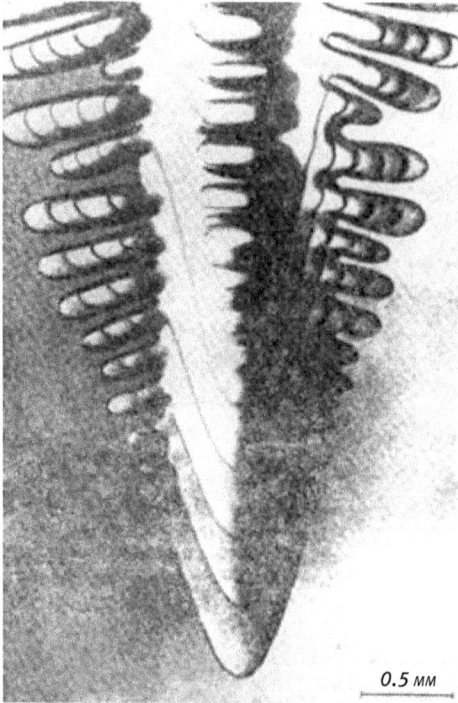

0.5 мм

Figure 3.9: The example of dendritic growth: successive growth stages for a dendritic succinonitrile crystal.

kinetics than that in the number of centers. The crystallization rate can be found by differentiation:

$$\frac{dV_t}{dt} = \frac{4}{3}\pi c^3 nt^3 V_0 \exp\left(-\frac{\pi}{3}c^3 nt^4\right).$$

Figure 3.10 presents the kinetic curves of crystallization of metals at different temperatures $(T_1 > T_2 > T_3 > T_4)$.

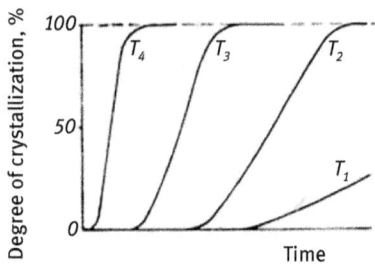

Figure 3.10: Kinetic curves of crystallization of metals at different temperatures $(T_1 > T_2 > T_3 > T_4)$.

As the crystallization temperature decreases (the degree of supercooling increases), the incubation period is reduced. The crystallization rate is highest in the middle part of the crystallization temperature range. Then it decreases due to the contact of the grown crystallized volumes. Figure 3.11 involves summary diagrams of the crystallization kinetics of metals (a) and alloys produced by rapid cooling to their amorphous state (b).

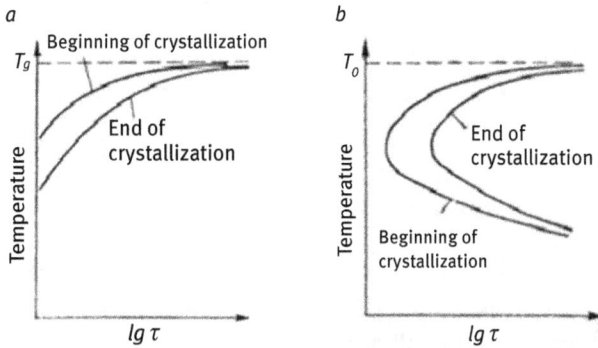

Figure 3.11: Summary diagrams of crystallization kinetics of metals (a) and alloys produced by rapid cooling to the amorphous state (b).

If one delays the crystallization process, the alloys when rapidly supercooled retain the structure of the liquid under solidification. In other words, they are in the amorphous state (see Section 3.9). The crystallization can be brought about with increasing temperature, in particular, under isothermal conditions. This case corresponds to the diagram in Figure 3.11b. For the overwhelming majority of metals and alloys, the crystallization process is impossible to delay by rapid cooling. Therefore, the bottom of the diagram cannot be realized, and amorphization does not occur (Figure 3.11a).

3.4.4 About grain size

To improve the mechanical properties of metals and alloys, fine grains must be produced in them. In many cases, such a procedure significantly enhances their performance in practice. Therefore, the grain size management becomes an important issue. Here, two main methods of grain grinding during crystallization should be indicated: an increase in the cooling rate and modification.

Every crystallization center independently formed gives growth to a crystallite (a grain) of a certain crystallographic orientation. With an increase in the cooling rate or the degree of supercooling, the number of crystallization centers increases (Figure 3.2). Consequently, the number of grains also rises. Obviously, in this case, the grain size diminishes. In the event of possibly reducing the growth rate of the centers when reaching the maximum value of the nucleation rate, the effect of grain grinding turns out to be maximal.

When modifying, various impurities contributing to heterogeneous nucleation are introduced into the melt. Crystallization centers artificially created can significantly grind the grain at the same cooling rate.

3.5 Dendritic growth

The growth of crystals in natural conditions occurs in accordance with the Gibbs–Curie–Wolf principle. It reads that the surface energy of the faces of a growing equilibrium crystal must be minimal:

$$\sum \sigma_i S_i = \min,$$

where σ_i is the surface tension of the ith face and S_i is its area. Usually, the faces with close-packed planes that have minimal surface energy grow rapidly. The Gibbs–Curie–Wolf principle is well satisfied at a small degree of supercooling. In this case, growing crystals acquire a correct face pattern (e.g., the growth of minerals in nature). For high supercooling, a transition occurs to another growth mechanism, a continuous one. Growing crystals often acquire a branched form or a dendritic structure.

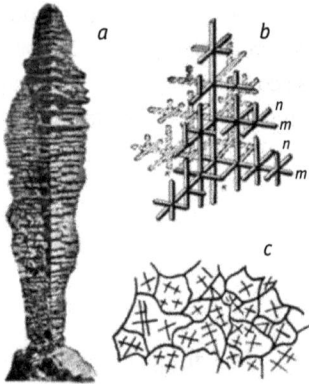

Figure 3.12: Chernov's dendritic crystal (a), Chernov's scheme for the growth of a dendritic crystal (b), and scheme of grains with dendrites (c).

Figure 3.12 presents a Chernov dendritic crystal about 40 cm long, taken out of the shrinkage shell of a 100-ton ingot, a Chernov's scheme for the growth of a dendritic crystal with the axes of the first, second, and third orders (k, m, n), and the resulting crystallized grains with orientation of original dendrites. Crystallization proceeds in the following sequence: first-order dendrite axes are formed, then higher order axes are created, and only after that, interdendritic spaces harden. The chemical composition of the dendritic axes and interdendritic spaces can differ dramatically (see

Section 3.7). This difference leads to heterogeneity of the structure and degrades the mechanical properties.

For explanation of the dendritic crystal growth, two theories are proposed: crystallographic and impurity ones. The former emanates from Chernov's position claiming that dendritic crystals (Chernov's discontinuous crystals) are elongated in the direction of the octahedral axes more quickly than along the other two axes (Figure 3.12b). The latter, the impurity theory, is based on the fact that convective fluxes in a liquid carry away impurities from the vertices and edges of the growing crystals, but on their faces, the impurity is precipitated and retards crystal growth (Bochvar). Therefore, the predominant growth occurs from the vertices and edges of the crystals. The dendritic impurity-rich gaps crystallize last. The interdendritic impurity-rich spaces crystallize last. Crystallites (grains) inherit a dendritic structure due to chemical heterogeneity (Figure 3.12c). A comprehensive analysis of dendritic growth refers to the special literature.

3.6 Concept of ingot structure

The cooling of liquid metal in the casting box occurs at a different rate at various points. This circumstance largely assigns the formation of the ingot structure. In the general case, the ingot structure consists of three zones such as a peripheral rapidly cooled layer, a zone of columnar crystals, and a central zone of equiaxed crystals.

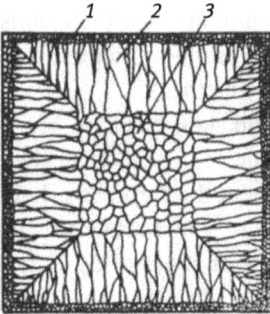

Figure 3.13: Cross section of an ingot (scheme): 1, a layer of fine grains; 2, a zone of columnar crystals; 3, a zone of equiaxed crystals.

Figure 3.13 demonstrates these zones in the cross section of the ingot. The peripheral zone involves small grains arisen from a large number of centers due to a high cooling rate and rapid heat dissipation to the casting box casting-form walls. Columnar crystals grow at a lower cooling rate in the direction perpendicular to the casting box casting-form walls, or, more precisely, in the direction opposite to the heat dissipation. The zone of columnar crystals can extend to the central part of the ingot. In this case, equiaxed crystals in the center part are absent (Figure 3.14a).

Levels of a flooded metal

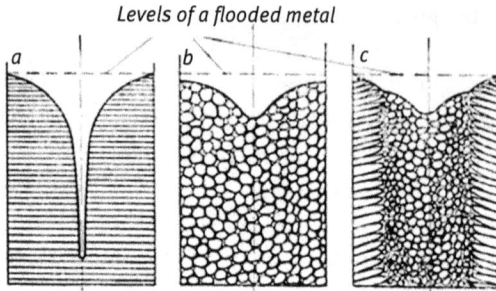

Figure 3.14: Longitudinal cross section of ingots (scheme):
(a) columnar crystals, (b) equiaxed grains, and (c) a mixed structure.

Columnar crystals have a crystallographic texture. In metals with a cubic lattice, they grow in the direction <100>. With the growth of columnar crystals, the degree of supercooling at the growth front decreases; and new randomly oriented centers arise in the central part of the ingot. Afterward, equiaxed crystals grow up from them. The zone of columnar crystals negatively affects the mechanical properties of a metal. So, the structure of equiaxed grains is usually desirable to obtain. The length of the zones and the ingot structure depend on many factors. Among these are the rate and direction of heat dissipation, the shape and material of the casting box casting form, convective fluxes during crystallization (mixing), properties of a crystallizing metal (thermal conductivity and heat capacity), the degree of overheating of the liquid metal relative to the melting temperature, the presence of dispersed inclusions (heterogeneous nucleation), and modification. By changing these factors, it is possible to control the formation of the structure of ingots (castings).

The vast majority of metals have the volume in the solid state less than in the liquid state. A decrease in volume during the crystallization of the ingot is called shrinkage. Shrinkage is manifested mainly in the formation of the shrinkage shell, as well as in the appearance of shrinkage porosity. Figure 3.14 shows schematically longitudinal cross sections of ingots with different lengths of zones of columnar and equiaxed crystals and various forms of a shrinkage shell. The shrinkage shell shape depends on the directions of heat dissipation. If heat removal prevails from the side casting box casting-form walls, the shrinkage shell spreads to a greater depth along the ingot axis (Figure 3.14a). Such a shrinkage shell shape is extremely unfavorable, and the discontinuity of a metal is maintained during hot deformation of the ingot. This defect adversely affects the mechanical properties of a metal.

3.7 Segregation

Segregation in an alloy is the heterogeneity of chemical composition in its different parts. Segregation can take place on a macroscopic scale in different zones of the ingot and can be observed on a microscale within each grain. In both cases, the reason for segregation to occur is that the equilibrium concentration cannot be achieved under crystallization. This is because of a decrease in temperature and a sufficiency of time. When segregated, the crystals formed are enriched with a refractory component, but a liquid is enriched with a low-melting constituent. Figure 3.15 illustrates this rule on the example of binary diagrams with unlimited solubility of the components.

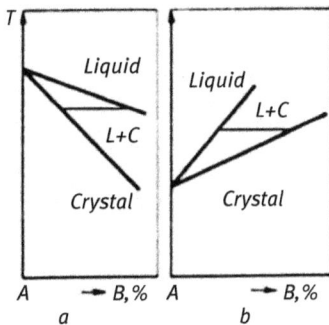

Figure 3.15: Parts of binary state diagrams explanting the reason for segregation.

The conoids in the diagrams show the equilibrium concentrations of coexisting phases (liquid and crystals) at a certain temperature. In real conditions, the equilibrium is established at a limited diffusion rate in the solid phase. Therein, the equilibrium concentration distribution between the phases cannot be achieved. As a result, once crystallized, the alloy maintains its chemical heterogeneity obtained during cooling in the two-phase region.

The central part of the ingot crystallizes last and is simultaneously enriched with a low-melting component. Therefore, there is a macroscopic heterogeneity of the chemical composition. Similarly, the interdendritic spaces crystallize last in each grain and are also enriched with the low-melting constituent. In such a way, a difference in the chemical composition appears between the axes of the dendrites and the interdendritic spaces. The heterogeneity of the chemical composition causes the heterogeneity of the structure and properties of the material. To eliminate the heterogeneity, homogenizing annealing must be performed and multiple recrystallizations should be carried out, when necessary (see Chapter 7). However, segregation is extremely difficult to completely get rid of in multicomponent alloys.

3.8 Eutectic crystallization

The eutectic crystallization is a special case. There are many industrial alloys based on eutectic-type systems. Eutectic alloys are important in the foundry industry since they have a low melting point (below the melting points of individual components), excellent casting properties, and a zero crystallization temperature range without segregation.

Eutectic crystallization is the simultaneous formation of two or more solid phases from a liquid. Bochvar was the first to put forward the theory of crystallization of eutectic alloys. In his experiments, he directly observed the crystallization of eutectics in the model system of organic substances such as azobenzene and piperonal. It is established that eutectic crystallization begins only under the condition of contact of crystals of various solid phase. As a result, colonies arise, often consisting of alternating plates of individual phases.

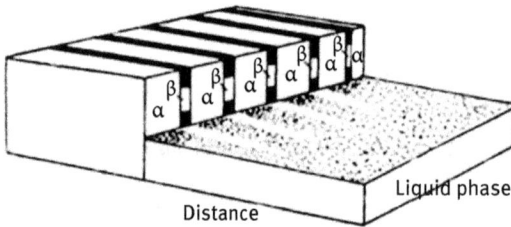

Figure 3.16: Scheme of the formation of a eutectic colony by Bunin.

Figure 3.16 illustrates the growth pattern of a eutectic colony consisting of plates of α- and β-phases. As can be seen, the concentrations are separated and the formation of the structure occurs at the front of the colony growth. With the colony growth, the diffusion paths are minimal; they are equal to half the sum of the thickness of the alternating plates. The concentration gradient in the liquid phase at the boundary of the growing colony is kept constantly high. Under these conditions, the growth rate of the eutectic colony is large, constant, and weakly dependent on concentration.

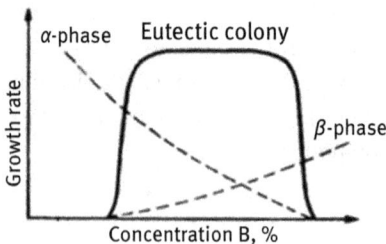

Figure 3.17: The linear growth rate of a eutectic colony and individual phases, depending on the concentration in an A–B binary system.

Figure 3.17 outlines schematically the dependencies of the growth rates of the eutectic colony and individual phases on concentration in the double-component system at a constant temperature. The colony growth rate is unchanged over a wide range of concentrations, and the individual crystals of each phase decreases as the content of the other component increases. The temperature dependencies of the growth rates of the eutectic and individual phases can be observed in Figure 3.18.

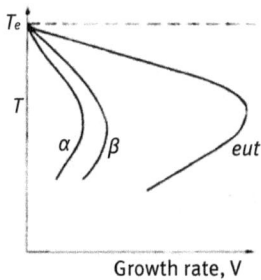

Figure 3.18: The linear growth rate of a eutectic colony and individual phases, depending on the temperature.

Two factors acting in opposite directions, the degree of supercooling and the diffusion mobility of atoms, are responsible for the run of all the curves. At a certain temperature, the growth rate reaches a maximum. At all temperatures, the growth rate of the eutectic is much greater than that of individual phases.

As mentioned earlier, the structure of a eutectic composition alloy usually consists of alternating plates of different phases. However, alloys in the concentration range near the eutectic can also have a eutectic or quasi-eutectic structure. This can be the case if the eutectic crystallization rate is greater than the crystallization rate of the individual phases. As an example, Figure 3.19 indicates the region of the possible formation of quasi-eutectic structures in the azobenzene–piperonal system.

Such a situation is inherent not only in eutectic metal systems but can also be extended to the case of a eutectoid (see, e.g., the pearlite transformation, Chapter 7).

3.9 Amorphous alloys

The amorphous state of alloys is also called metallic glass or frozen liquid. These terms reflect the main property of an amorphous alloy, such as the absence of a crystalline structure. Amorphous Au–Si and Au–Ge alloys produced by rapid quenching of the melt became a starting point in intensive studies of amorphous alloys (P. Duvez, 1960). The condition for obtaining an amorphous state is a high cooling rate (see Figure 3.11b). For metal alloy systems being of industrial interest, cooling rates of 10^5–10^6 K/s are required. Such rates are achieved by means of special methods: by cooling a jet or drops of the melt on a cold surface of a rotating copper

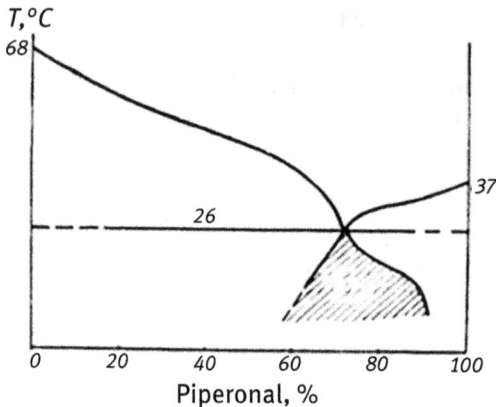

Figure 3.19: The region of quasi-eutectic structures (hatched) in the azobenzene–piperonal system (by Bochvar).

cylinder, by flattening a drop of the melt by cooling plates, and by rolling the melt in cooling rolls. In all cases, amorphous alloys are obtained in the form of a thin ribbon or thin "flakes."

An amorphous state can be obtained only for certain groups of alloys. The main group of alloys contains a sufficient amount of nonmetallic atoms (B, Si, P, and C). Such atoms with a small atomic radius are believed to occupy the voids of the atomic packing of the melt, delaying the formation of a crystalline structure. The other group is low-eutectic temperature alloys ("deep" eutectic). For example, in Au–Si systems, the melting points of gold, silicon, and eutectics are 1,064, 1,415, and 370 °C, respectively. The melt has a high viscosity at low temperatures, which impedes crystallization. Finally, if the crystalline phase of an alloy has a complex lattice that requires considerable repacking of atoms, such a lattice can be difficult to construct. In this case, the alloy solidifies when cooled, saving the structure of a liquid. The solidification temperature of amorphous alloys (glass transition temperature) is not constant. It varies over a wide range of $(0.35–0.60)$ T_0, where T_0 is the equilibrium temperature of the melt and crystals (see Figure 3.1).

In an amorphous state, no long-range order in the arrangement of atoms exists. However, the short-range order remains as in a liquid. Regardless of the type of atoms, their spatial arrangement is characterized by topological short-range order (TSO). The order in the arrangement of atoms of each type is characterized by chemical short-range order (CSRO). The short-range order in an amorphous alloy can be described by the radial distribution function (RDF). The latter serves as a measure of the probability of finding an atom at a distance r from the middle central atom. RDF can be determined experimentally using scattering curves of X-rays, neutrons, and, less accurately, electrons. As an example, Figure 3.20 presents the dependence of the X-ray scattering intensity on the scattering angle for an amorphous $Ni_{81}P_{19}$ alloy.

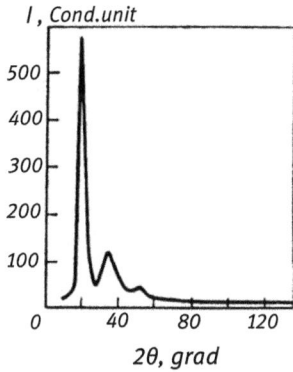

Figure 3.20: X-ray scattering intensities depending on the scattering angle for an amorphous $Ni_{81}P_{19}$ alloy.

Analyzing such dependencies yields the RDF. Different models of the amorphous state underlie the theoretical definition of RDF. For example, the Bernal model offers an amorphous alloy as random packing hard spheres in a tight mobile ("rubber") shell. Further, theoretical RDFs are compared with those found from experiments. Available information on the analysis of various models of the amorphous state is present in the special literature.

An important characteristic of an amorphous alloy is free volume. Figure 3.21 shows schematically the change in specific volume during the glass transition and crystallization processes.

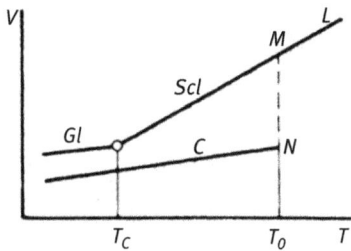

Figure 3.21: Change in the specific volume during the glass transition and crystallization: L, a liquid; Scl, a supercooling liquid; Gl, glass; C, a crystal; MN, a decrease in volume upon crystallization.

The alloy decreases its specific volume abruptly when crystallized but continuously upon transition to the amorphous state. The specific volume in the amorphous state is always greater than in the crystalline state; this difference is called free volume. The latter's structure can be represented either as a uniformly distributed empty space or as a concentrated one in the form of dispersed pores whose size does not exceed the size of the smallest atoms of the alloy. For a number of amorphous alloys, the value of the relative free volume is in the range of 0.025–0.030.

An amorphous state is metastable, and its stability is due to the low diffusion mobility of atoms. Heating or prolonged holding causes a change in the structural state and properties of an amorphous alloy even before its crystallization. Such changes are referred to as relaxation. These changes can persist after cooling (irreversible relaxation) or return to the initial state (reversible relaxation). Examples of changes in length (irreversible relaxation) and Young's modulus (reversible relaxation) when two amorphous alloys are heated can be observed in Figure 3.22.

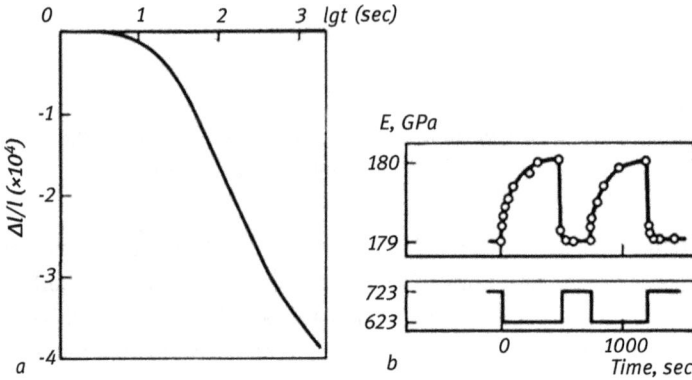

Figure 3.22: An irreversible decrease in length of an amorphous $Fe_{40}Ni_{40}B_{20}$ alloy during isothermal holding at 623 K (a) and short-term reversible changes of Young's modulus in a $Co_{58}Fe_5Ni_{10}B_{16}Si_{11}$ alloy between 623 and 723 K (b).

The occurrence of irreversible relaxation is easier to explain than the causes of reversible one. The former is associated with a change in TSO; the latter may be related to a reversible change in CSRO as TSO is left unchanged.

When heated up to temperatures in the range of $(0.40–0.65) T_0$, amorphous alloys crystallize. To study this process, differential scanning calorimetry (DSC) and transmission electron microscopy (TEM) are utilized as the most informative methods. DSC curves need to detect the crystallization temperature. TEM method and X-ray analysis allow one to identify the crystalline phases formed. Crystallization is often brought about in one stage but may be divided into a few stages.

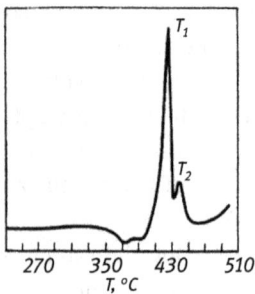

Figure 3.23: A DSC crystallization curve for an amorphous $Pd_{77.5}Cu_{6.0}Si_{16.5}$ alloy during heating at a rate of 20 K/min. The ordinate axis designates liberated energy (an exothermic effect) or absorbed energy (an endothermic effect).

As shown in Figure 3.23, the DSC curve for an amorphous $Pd_{77.5}Cu_6Si_{16.5}$ alloy has two crystallization peaks, T_1 and T_2. The first peak is due to the primary crystallization that modifies the composition of the amorphous matrix. The second peak is given rise to the formation of another phase at a higher temperature. Typically, the crystallization peaks are located close to each other. During crystallization, the crystalline phase most often forms the same chemical composition as the amorphous one. However, eutectic decomposition into two crystalline phases is also possible to happen. In this case, eutectic crystallization takes place in the solid state during heating rather than during cooling.

Amorphous alloys have a great advantage as compared to crystalline ones due to a high degree of chemical homogeneity. They consist of one phase because there is no segregation. The alloys contain no grain boundaries, dislocations, and vacancies. These features of the structure dictate the lack of texture and anisotropy of properties, high strength, corrosion resistance, high magnetic permeability, and low energy loss for magnetization reversal. Amorphous alloys of FeBSi and FeBSiC systems are applied for the manufacture of transformer cores. An FeCoSiB alloy is used for producing magnetic recording heads. There is another possibility of using the amorphous state: it is the starting point for obtaining a submicrocrystalline and nanocrystalline structure upon heating. Such a structure possesses high mechanical and other properties (see Section 5.4).

In conclusion, it should be noted that in solitary instances, an amorphous state can be obtained by high-degree plastic deformation. For example, a TiNi shape-memory alloy becomes amorphous after deformation in Bridgman anvils (see Section 5.4).

Chapter 4
Diffusion in metals

Diffusion in a system is usually called the process of the spontaneous motion of atoms for equalizing their concentrations. However, this definition is not accurate. A driving force for the diffusion transfer is the difference in chemical potentials at different points of the system rather than the difference in concentrations. As an example, uphill diffusion (diffusion in a stress field) refers to a process of increasing the concentration difference. Another example is given in Section 4.8, when describing the Darken experiment. Despite this, in many cases, the description of diffusion processes may leave aside the concept of chemical potential.

Diffusion processes play an important role in physical metallurgy. Firstly, diffusion takes place in many high-temperature phase transformations and structural changes such as decomposition of supersaturated solid solutions, coagulation, recrystallization, homogenization, creep, oxidation, chemical heat treatment, and sintering. Secondly, diffusion is the main tool for studying crystal structure defects, for example, point defects and grain boundaries.

4.1 Phenomenological theory of diffusion: Fick's laws

A diffusion flux is described by the same differential equation as a heat flux in the case of thermal conductivity. In one case, this is the transfer of heat; in the other case, it is the transfer of atoms of a dissolved substance. The one-dimensional heat equation has the form

$$dQ = -\lambda S\ (\partial T / \partial x)\ dt,$$

where dQ is the heat flux through the area S over a period of time dt, λ is the coefficient of thermal conductivity, and $\partial T / \partial x$ is the temperature gradient. By analogy, the diffusion equation appears as follows:

$$dq = -DS\ (\partial c / \partial x) dt,$$

where dq is the substance flux through the area S over a period of time dt, D is the diffusion coefficient, and $\partial c / \partial x$ is the concentration gradient. Let us introduce the designation for the unit flux:

$$J = dq /\ S\ dt.$$

In the one-dimensional case, the first law of diffusion (Fick's first law) takes the form

$$J = -D\ (\partial c / \partial x).$$

https://doi.org/10.1515/9783110758023-004

The sign "–" (minus) stands for the flux direction toward a decrease in concentration. The flux disappears in the event of fulfilling the condition $\partial c/\partial x = 0$. D is a constant that depends on the nature of the solvent and the solute. Dimension D is expressed as "cm^2/s." In the SI system, it has the unit meter squared per second (m^2/s).

For the three-dimensional case, the first diffusion law is written as follows:

$$J = -D\nabla c,$$

where ∇c is the concentration gradient, $\nabla c = \partial c/\partial x + \partial c/\partial y + \partial c/\partial z$.

If a steady state cannot be achieved, that is, if the concentration of a solute is time dependent at each point, another equation is used. The latter is derived from the first diffusion law and the substance balance condition that reads that the amount of a diffusing substance remains constant, and the substance is only redistributed. Simple transformations lead to the second law of diffusion (Fick's second law). For the one-dimensional case, it has the form

$$\partial c/\partial t = D\ (\partial^2 c/\partial x^2),$$

where $\partial c/\partial t$ is the rate of change in concentration. The equation describes the change in the concentration of a diffusing substance in space and in time.

For the three-dimensional case, the second diffusion law can be stated as follows:

$$\partial c/\partial t = D\ (\partial^2 c/\partial x^2 + \partial^2 c/\partial y^2 + \partial^2 c/\partial z^2) \quad \text{or}$$
$$\partial c/\partial t = D\Delta c,$$

where Δ is the Laplace operator. Another form of writing Fick's second law is

$$\partial c/\partial t = -\operatorname{div}J,$$

where div J is the flux divergence.

4.2 The diffusion coefficient

The diffusion coefficient D is a parameter that characterizes the diffusion rate. For metals, with increasing temperature, the diffusion coefficient increases dramatically, changing according to the exponential law:

$$D = D_0 \exp(-Q/RT),$$

where D_0 is the pre-exponential factor, Q is the activation energy, R is the gas constant, and T is the absolute temperature. The Q and D_0 quantities are related to the physicochemical properties of a solvent metal and a solute. Table 4.1 lists the values of Q and D_0 for some frequently encountered diffusion pairs.

As shown in Table 4.1, the activation energy of diffusion of carbon in α-Fe is small as compared to that of self-diffusion of iron. The diffusion activation energy

Table 4.1: Diffusion constants included in the equation $D = D_0 \exp(-Q/RT)$.

Solvent	Diffusing element	D_0 (cm²/s)	Q (cal/g-atom)
HCC γ-Fe	Fe (self-diffusion)	0.6	68,000
BCC α-Fe	Fe (self-diffusion)	5.8	59,700
BCC α-Fe	C	0.02	20,000
BCC α-Fe	N	4.6×10^{-4}	17,950
HCC Ni	Ni (self-diffusion)	1.3	66,800
HCC Cu	Cu (self-diffusion)	0.2	47,100
BCC W	W (self-diffusion)	11.5	142,000

of nitrogen in α-Fe is even lower. Consequently, the diffusion rate of carbon and especially of nitrogen is greater than that of iron. The self-diffusion activation energy of iron in a close-packed face-centered cubic (FCC) γ-Fe lattice is higher than that in a body-centered cubic (BCC) α-Fe lattice. Accordingly, the self-diffusion rate of iron in the BCC lattice is greater than in the FCC one. It can be noticed that the self-diffusion activation energy of high-melting tungsten is substantially higher than the self-diffusion activation energy of lower melting iron, nickel, and copper.

4.3 Examples of solving diffusion problems

Diffusion problems can be divided into two groups. The first group involves calculating the diffusion coefficient D, the pre-exponential factor D_0, and the activation energy Q upon measuring diffusion characteristics.

By the preassigned diffusion coefficient D, the second group determines diffusion characteristics such as the distribution of concentration of a diffusing substance and the amount of a substance penetrated (or left) into the material for a certain time.

To solve the problems posted, it is necessary to know the analytical dependencies between the diffusion characteristics in the integral form. These dependencies can be derived by integrating the differential equation of the second diffusion law under initial and boundary conditions of a specific physical problem. Let us indicate two methods for solving the differential diffusion equation (the heat conductivity equation). The former is that of separating variables (also known as the Fourier method), and the latter represents the source method or the Green's function method.

Suppose that the initial distribution of concentrations is given in the form

$$c(x,0) = f(x)$$

and the diffusion coefficient D is concentration independent. The solution of equation

$$\partial c / \partial t = \partial / \partial x (D \times \partial c / \partial x)$$

in the general case, for an unlimited body can be found as follows:

$$C(x, t) = T(t) \times X(x),$$

where $T(t)$ depends only on time and $X(x)$ depends only on the coordinate.

The general solution of the diffusion equation is rarely used, but it produces particular solutions. Let us dwell on some of them.

1. Solution for bidirectional diffusion of atoms from an infinitely thin layer inside a body. The distribution of the concentrations is

$$c(x, t) = \frac{q}{2\sqrt{\pi Dt}} \exp\left(-\frac{x^2}{4Dt}\right),$$

where c is the concentration (the amount of a substance per 1 cm^3) and q is the amount of a substance per 1 cm^2 of the layer.

The function $\exp(-(x^2/(4Dt)))$ is even; therefore, the concentration distribution is symmetrical relative to the middle of the layer ($x = 0$) (Figure 4.1). The initial concentration distribution "blurs" over time. The maximum value, $c_{max} = \left(q / \left(2\sqrt{\pi Dt}\right)\right)$, decreases inversely proportionally to $\sqrt{t}$ at $x = 0$.

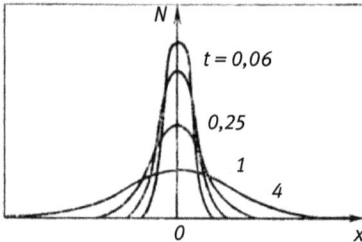

Figure 4.1: Concentration distribution in diffusion from an infinitely thin layer into an unlimited size body.

2. Solution for diffusion from a layer of finite thickness. Its thickness is 2 h.

$$c(x, t) = \frac{c_0}{2}\left[\text{erf}\frac{h+x}{2\sqrt{Dt}} + \text{erf}\frac{h-x}{2\sqrt{Dt}}\right].$$

Here, c_0 is the initial concentration; erf is the Gauss error function:

$$\text{erf } y = \frac{2}{\sqrt{\pi}} \int_0^y \exp(-z^2)\, dz.$$

As is seen from the graph of the function $c(x, t)$ in Figure 4.2, the concentration vanishes everywhere (erf $0 = 0$) as $t \to \infty$ because a finite amount of a substance $q = 2c_0h$ is distributed over an infinite region.

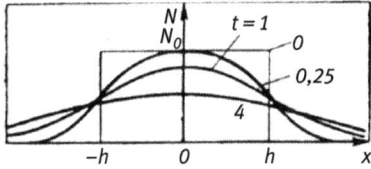

Figure 4.2: Concentration distribution in diffusion from a layer of finite thickness.

3. Solution for diffusion from semi-infinite space.
The initial distribution of concentrations is

$$c(x, 0) = c_0 \text{ for } x < 0,$$

$$c(x, 0) = 0 \text{ for } x > 0.$$

The solution is

$$c(x, t) = \frac{c_0}{2}\left(1 - \text{erf}\frac{x}{2\sqrt{Dt}}\right).$$

For all $t > 0$, the concentration in the separation plane $(x = 0)$ is constant and equal to $c_0/2$ (Figure 4.3).

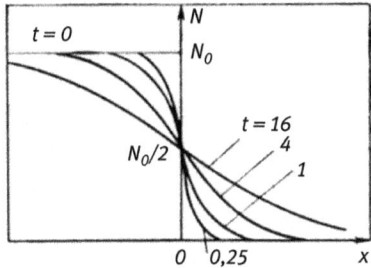

Figure 4.3: Concentration distribution in diffusion from semi-infinite space.

4.3.1 Evaluation of diffusion paths

For an approximate estimate of diffusion paths, the following relation is utilized:

$$x \sim \sqrt{Dt}.$$

In the problem of diffusion from an infinitely thin layer, we determine which law is responsible for changing the distance to the plane where the concentration is "e" times less than that in the plane $x = 0$. For $x = 0$, we have

$$c(0,t) = \frac{q}{2\sqrt{\pi Dt}} \exp(0).$$

In the plane, where the concentration is "e" times less, $c(x,t) = (c(0,t)/e)$.

Now, we set the concentrations in both the planes equal to each other:

$$\frac{c(0,t)}{e} = \frac{q}{2\sqrt{\pi Dt}} \exp\left(-\frac{x_1^2}{4Dt}\right); \quad e^{\frac{x_1^2}{4Dt}}e = 1; \quad -\frac{x_1^2}{4Dt} + 1 = 0; \quad x_1^2 = 4Dt; \quad x_1 = 2\sqrt{Dt}.$$

Thus, the distance to the plane varies proportionally to $\sqrt{t}$.

4.4 Diffusion in a stress field

The cases dealt with diffusion in the previous section refer to diffusion due to a concentration gradient. In the general case, the flux of a substance is the sum of the fluxes exposed to a concentration gradient and an additional force field. Such a field can be a field of inhomogeneous stresses (deformations). Deformations can be provoked by embedded atoms or vacancies. The stress field around the embedded atom in a solid solution contributes to the attraction of the atom to a dislocation. It may turn out that the total diffusion flux is directed toward a higher concentration. This is the case of uphill diffusion mentioned above.

The second diffusion equation that allows for the stress field has the form:

$$\frac{\partial c}{\partial t} = D'\frac{\partial^2 c}{\partial x^2} - D''\frac{\partial^2 \varepsilon}{\partial x^2}.$$

Here ε is the elastic deformation, and the diffusion coefficient D'' characterizes the flux associated only with the stress (strain) gradient. The value of D'' is proportional to the relative difference of the atomic radii of the alloy components, $(r_B - r_A)/r_A$ (by Konobeevsky). The elastic deformation causes large-radius and small-radius atoms to travel into the stretched and compressed layers, respectively.

If the initial components of an alloy are statistically uniformly distributed in the crystal lattice, diffusion, described by the second term, proceeds under a stress field. The uphill diffusion results in separating the components and in arising chemical heterogeneity. Described by the first term, a counter diffusion flux emerges to equalize the concentrations. With a constant deformation ε, an equilibrium distribution of the components will occur when both fluxes become equal:

$$D'\frac{\partial^2 c}{\partial x^2} = D''\frac{\partial^2 \varepsilon}{\partial x^2}.$$

4.5 Atomic diffusion theory: diffusion coefficient

Observing under a microscope the Brownian motion of any microparticles in fluids, it can be asked how random movements of these microparticles are related to their macroscopic displacement. Marjan Smoluhovsky and Albert Einstein were the first to formulate this issue known currently as the problem of random walks. They succeeded in establishing a relationship between the length of jumps of a particle and its jump frequency during Brownian motion on the one hand, and the diffusion coefficient, on the other hand. It turned out that diffusion occurs very slowly $(D \approx 10^{-8} \text{cm}^2/\text{s})$ if there are no external impacts. It may seem that we have left aside diffusion in crystalline solids but this is not the case. Diffusion in crystals exhibits as periodic atomic jumps from one lattice site to another. Mathematical analysis of the random-walk problem as applied to diffusion in crystals allows one to relate to the observed macroscopic diffusion coefficients and the frequency and length of jumps of diffusing atoms. It is worth noting that diffusion experiments can be used not only for studying actual diffusion processes themselves, but also for examining interatomic interactions, as well as crystalline lattice defects.

4.5.1 Random movements and diffusion coefficient

Let us look at a simple one-dimensional model used for deriving an approximate relationship between the diffusion coefficient D, the frequency, and the length of jumps of atoms. Suppose that the crystal sample has a concentration gradient along the x-axis (Figure 4.4).

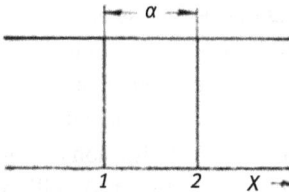

Figure 4.4: Illustration of the first diffusion law.

Here, 1 and 2 are adjacent crystal planes, and the distance between them is equal to α or to the jump length. We designate the number of atoms diffusing per unit area of plane 1 and plane 2 as n_1 and n_2, respectively. Each atom jumps on an average Γ times per second. The number of atoms leaving the plane 1 in a short time period of δt is $n_1 \Gamma \delta t$. Half of these atoms come into plane 2, and the other half fall into the adjacent plane to the left. Therefore, the number of atoms jumping from plane 1 to plane 2 over a time δt can be written as follows:

$$1/2\,n_1\,\Gamma\,\delta t.$$

½ n_2 δt atoms jump from plane 2 to plane 1 over time δt. The total flux from plane 1 to plane 2 per unit time is

$$J = \frac{\text{Number of atoms}}{(\text{Area})(\text{Time})} = 1/2(n_1 - n_2)\Gamma.$$

The difference $(n_1 - n_2)$ can be related to the concentration or number of atoms per unit volume because $(n_1/\alpha) = c_1$ and $(n_2/\alpha) = c_2$. Therefore, we arrive at the following formula:

$$J = 1/2(c_1 - c_2)\alpha\Gamma.$$

As diffusion proceeds, the concentration c changes rather slowly depending on the composition, so that

$$c_1 - c_2 = -\alpha\frac{\partial c}{\partial x},$$

$$J = -\frac{1}{2}\alpha^2\Gamma\frac{\partial c}{\partial x}.$$

We have deduced the first law of diffusion, where

$$D = \frac{1}{2}\alpha^2\Gamma.$$

We have taken for the jump frequency Γ to be the same in planes 1 and 2 for jumps to the left and to the right. The flux from plane 1 to plane 2 arises because $n_1 > n_2$. It is reasonable to compare the value of α with the value of the interatomic distance in the lattice, $\alpha \sim 1$ Å. Then, the preassigned diffusion coefficient D makes it possible to estimate the value of Γ. For carbon in α-Fe, the diffusion coefficient is $D \approx 10^{-6}$ cm^2/s at 900 °C. If $\alpha \approx 10^{-8}$ cm, then $\Gamma \approx 10^{10}\text{s}^{-1}$. In other words, the carbon atom changes its position 10 billion times a second. Beneath the melting point, most metals with FCC and hexagonal close-packed lattices have $D \approx 10^{-8}$ cm^2/s (D is the self-diffusion coefficient). For $\alpha \approx 10^{-8}$ cm, we get $\Gamma \approx 10^8$ s^{-1}. Atoms change their position 100 million times per second. This number may seem to be too large. However, it is not hard to recollect that the average frequency of atomic vibrations (the Debye frequency) is 10^{12}–10^{13} s^{-1} so that the atom changes its position only once per 10^4–10^5 vibrations. Thus, even beneath the melting point, an atom vibrates around its rest position in a crystal for a long time. It is also useful to compare the displacement of an atom from its initial position during diffusion (diffusion path, L_D) with the total distance that an atom travels for the same time (L). With diffusion for 100 hours, we come up with $L \approx 1$ km, $L_D \approx 0.6$ mm. That is, passing the total distance of 1 km, the atom shifts from its initial position by less than 1 mm. This estimate shows how small the efficiency of random walks (random motion) is for moving a diffusing substance in the process of diffusion annealing.

4.6 Atomic diffusion mechanisms

Atomic diffusion mechanisms deal with the various paths of atoms to move, resulting in diffusion.

4.6.1 An interstitial mechanism

An interstitial mechanism is realized when an atom moves from an interstitial lattice site to a neighboring one. Figure 4.5 shows the path of an embedded atom diffusing by the interstitial mechanism in the (100) plane of the FCC lattice.

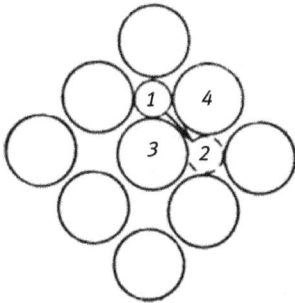

Figure 4.5: Path of an embedded atom in the (100) plane of the FCC lattice under diffusion by an interstitial mechanism.

The atom moves from position 1 to position 2. For this, atoms 3 and 4 must "move apart" to make way for the diffusing atom. Otherwise speaking, before the atom will jump, a noticeable local expansion should occur, which requires spending energy. Consequently, the diffusing atom must overcome the energy barrier that arose. Figure 4.6 illustrates the FCC-lattice interstitials or the inclusion positions denoted by the crosses.

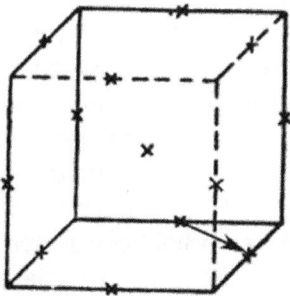

Figure 4.6: Inclusion positions (crosses) in the unit FCC cell.

These interstitials themselves constitute the FCC lattice. The atom diffuses via the interstitial mechanism, jumping from one position of the interstitial lattice to another.

Such a mechanism is thought to take place in diffusion of carbon and other embedment atoms in the α- and γ-iron. However, in the event of creating significant lattice distortions by the atoms diffusing through the interstitial mechanism (the sizes of the solvent and solute atoms are close), this mechanism is not realized.

4.6.2 The vacancy diffusion mechanism

In all crystals, some lattice nodes are empty. These nodes are vacancies. If one of the neighboring atoms occupies a vacancy, it diffuses by a vacancy mechanism. The FCC-lattice vacancies where the atom strives to are shown in Figures 4.7 and 4.8.

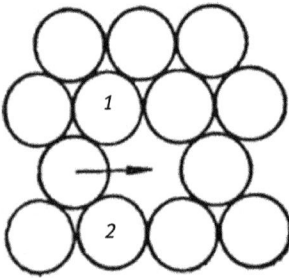

Figure 4.7: Vacancy diffusion mechanism in the (111) plane of the FCC lattice.

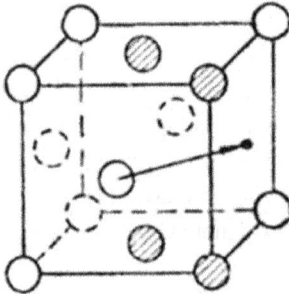

Figure 4.8: Vacancy diffusion mechanism in the unit cell of the FCC lattice.

In order for the diffusing atom to travel, atoms 1 and 2 should "move apart" (see Figure 4.7). The scheme in Figure 4.8 depicts these atoms as shaded. The nature of the obstacles to diffusion via the vacancy mechanism can be explained in such a way.

The lattice distortion energy due to diffusion of iron atoms by a vacancy mechanism is approximately equal to that caused by diffusion of carbon atoms by an interstitial mechanism in the same FCC lattice. However, carbon diffuses much faster than iron. This is because there are many vacancies for carbon atoms (neighboring interstitials) and very few ones for iron atoms. In other words, one needs to "wait" for the vacancy to come.

The vacancy diffusion mechanism prevails in FCC metals and alloys and operates in BCC and hexagonal metals, as well as in ionic crystals and oxides.

4.6.3 Diffusion along interstitials by a replacement mechanism: a clustering mechanism (crowdion)

The size of impurity atoms forming interstitial solid solutions is noticeably smaller than that of solvent metal atoms. As said earlier, the impurity atoms diffuse through the interstitial mechanism. For an embedded large-size atom to diffuse, there exist special diffusion mechanisms along the interstitial sites, namely, interstitial–substitutional mechanisms. Figure 4.9 demonstrates a mere example of pushing out the nearest neighbor into the interstitial lattice site by the embedded atom followed by substituting for the former.

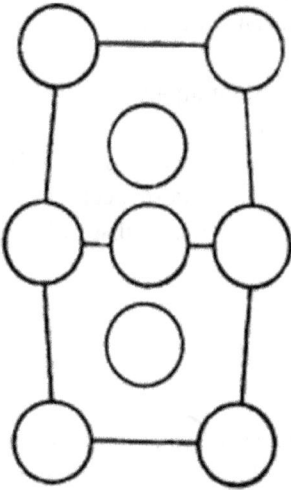

Figure 4.9: An atom in the interstitial site in the (100) plane of the FCC lattice.

As an instance, it is this mechanism that makes silver diffuse in AgBr. A silver ion is smaller than a bromine ion; therefore, the embedded silver ion does not distort the lattice. With an identical size of atoms (e.g., during copper self-diffusion), the configuration shown in Figure 4.10 has a lower distortion energy than that in Figure 4.9.

Here, two atoms share "one place" in the (100) plane of the FCC lattice. Diffusion can proceed when the pair is rotated, or when one of the atoms is moved into the neighboring cell.

The crowdion diffusion mechanism (a clustering mechanism) can be viewed in Figure 4.11.

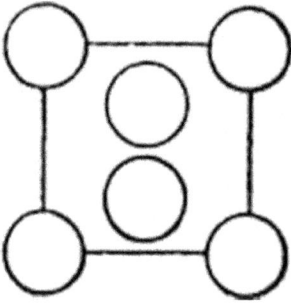

Figure 4.10: Two atoms share "one place" in the (100) plane of the FCC lattice.

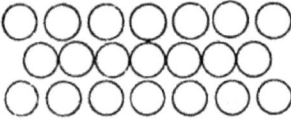

Figure 4.11: An atom cluster in the (111) plane of the FCC lattice (the middle row).

In the middle row on the (111) plane of the FCC lattice, an extra atom appears in the direction of the dense packing. In the row, a displacement similar to the dislocation one is possible to emerge. The energy of such a displacement is small, and thus diffusion takes place.

In metals with a close-packed lattice, the energy required for an atom to migrate to an interstitial site is so high that the concentration of interstitial atoms in an annealed sample is negligible. However, a large concentration of these atoms can be achieved by bombarding the samples with high-energy particles or during cold deformation. In such samples, diffusion can occur involving interstitial–substitutional mechanisms.

4.6.4 The ring diffusion mechanism

It should be mentioned that there is yet another diffusion mechanism, the ring diffusion one. Figure 4.12 represents the cases of exchanging places of two (simple exchange) and three neighboring atoms (cyclic exchange). The latter creates distortions in the lattice much less than the former. Apparently, such a diffusion mechanism occurs in metals with a loose packing of atoms (e.g., in metals with a BCC lattice).

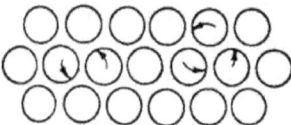

Figure 4.12: Ring diffusion mechanism.

Thus, vacancy and interstitial diffusion mechanisms are most often implemented in metals and alloys. The vacancy mechanism takes place in substitutional solid solutions when the sizes of different atoms differ little. The interstitial mechanism is inherent to embedded atoms.

4.7 Paths of accelerated diffusion

Disturbance of the order in the arrangement of atoms in an ideal crystal lattice must facilitate the motion of atoms, hence, accelerate diffusion. Therefore, defective elements of the structure (vacancies, dislocations, grain boundaries, and surface) are called "paths of accelerated diffusion."

Vacancies are an equilibrium defect. The free energy of a crystal (as a function of the concentration of vacancies) is minimal, provided the concentration of vacancies is thermodynamically equilibrium. The concentration of equilibrium vacancies is relatively small, but it increases with increasing temperature. Therefore, for metals and alloys, the vacancy concentration is possible to achieve substantially higher than the equilibrium one. The procedure implies many different ways such as high-temperature quenching, irradiation, cold deformation, and phase transformations. Excessive vacancies can sink to grain boundaries, to dislocations, to the outside surface of the samples, and annihilate. However, they can be partially saved, increasing the free energy of an alloy.

Dislocations and grain boundaries are nonequilibrium defects. Due to them, the free energy of a crystal increases. Therefore, processes taking place in the crystal must reduce the dislocation density and the total area of the grain boundaries. However, for the processes to be noticeable, the following conditions are necessary: sufficient driving forces and sufficient mobility of the defects themselves. Otherwise, these defects remain "frozen." Let us consider the influence of the defects on diffusion. Among these are zero-dimensional defects as excessive vacancies, one-dimensional defects as dislocations, two-dimensional defects as grain boundaries, and three-dimensional defects as the outer surface of a crystal.

4.7.1 Nonequilibrium (excessive) vacancies

The equilibrium concentration of vacancies in copper amounts to $\sim 10^{-5}$ and $\sim 10^{-19}$ at 1,000 °C and at room temperature, respectively. It is worth pointing out that the vacancy concentration dependence on temperature is exponential. If one succeeds in being preserved all vacancies in copper by quenching, the vacancy concentration would be 10^{14} times more than the equilibrium one. However, in reality, a significant part of the vacancies disappears during cooling, but the excess remains and is several orders of magnitude.

The concentration of vacancies is included in the expression for the self-diffusion coefficient D:

$$D = D_v N_v,$$

where D_v is the diffusion coefficient of vacancies and N_v is their atomic fraction. Thus, lattice oversaturation with vacancies accelerates diffusion by a vacancy mechanism. It is important to note that the average lifetime of vacancies should be large enough. This condition is satisfied if the density of sinks is low and if the path of vacancies to a sink is not too small. Usually, the grain size in an alloy or the sample size should be at least 10^{-2} cm.

4.7.2 Dislocations

There is a well-known fact that diffusion becomes faster under the influence of plastic deformation. In a well-annealed single crystal, the dislocation density is 10^4–10^6 cm^{-2}. Plastic deformation increases the dislocation density to 10^{11}–10^{13} cm^{-2}. The density of vacancies also increases but their lifetime at such a density of sinks is very small. Then, in this case, the main role in accelerating diffusion is played by dislocations. The following conclusions on the dislocation influence on diffusion can be made. Dislocations serve as a transport highway of accelerated motion for atoms to migrate. The self-diffusion coefficient along dislocations is several orders of magnitude greater than that in the crystal lattice. The greater the relative contribution of diffusion along dislocations, the lower the temperature.

Experiments allow one to determine not the diffusion coefficient along dislocations D_d, but the product

$$f_d D_d,$$

where f_d is the cross section of the dislocation tube in the form of a narrow cylinder. The diffusion coefficient in the tube is much greater than in volume. The coefficient D_d depends on the structure of the dislocation core, determined by the Burgers vector and the direction of the dislocation line. Calculating the diffusion coefficient D_d meets fundamental difficulties. Firstly, diffusion fluxes along the dislocations and in the dislocation-free volume need to be separated. Secondly, the cross-sectional size of the tube is never exactly known, but rather it is chosen arbitrarily. By making some admissions, these difficulties can be circumvented.

The estimate shows that the dislocations make a contribution, ranging from very small to significant, to the total diffusion flux, depending on the density of dislocations and the temperature of a diffusion experiment. So, for silver, the value of D_d is only 0.08 of the total flux at a temperature of 590 °C (0.7 T_{melt}) and a dislocation density $\rho = 10^7$ cm^{-2}. At 220 °C (0.4 T_{melt}) and at the same dislocation density, D_d rises to 10^3. Therefore, at this temperature, the entire flux is governed by diffusion along the dislocations. The activation energy of dislocation diffusion turned out to be significantly lower than that of bulk diffusion. For example, for self-diffusion of silver, the corresponding numbers are 20,000 and 44,000 cal/g-atom. Therefore, the contribution of dislocation diffusion increases with decreasing temperature. This is evident from the above example.

4.7.3 Grain boundaries

Let us take a logarithm of the expression for the diffusion coefficient:

$$D = D_0 \exp\left(- Q/RT\right).$$

Then, we obtain

$$\ln D = \ln D_0 - \frac{Q}{RT}.$$

The function $\ln D = f(1/T)$ is depicted as a straight line if Q and D_0 are constants. The activation energy Q is calculated from the slope of the line:

$$\frac{d \ln D}{d\left(\frac{1}{T}\right)} = - \frac{Q}{R}.$$

The segment along the y-axis for $1/T = 0$ is equal to $\ln D_0$. However, for a fine-grain polycrystalline sample, the graph of the function $\ln D = f(1/T)$ consists of two segments (two straight lines) (Figure 4.13).

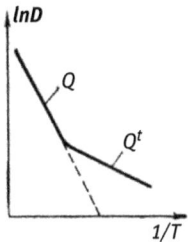

Figure 4.13: The function $\ln D = f(1/T)$ for a fine-grain polycrystalline sample.

A transition from one segment to another is observed at temperatures (0.6–0.7) T_{melt}. The high-temperature part of the plot describes bulk diffusion; the low-temperature part involves diffusion along the grain boundaries. The coefficient (D') of the boundary diffusion is greater than that (D) of the bulk one. The activation energy of the boundary diffusion is lower than that of the bulk diffusion ($Q' < Q$).

In investigating boundary diffusion, the issue of the separation of diffusion fluxes over grain boundaries and volume should be raised. Fisher proposed a mathematical model that takes into account the rapid diffusion process along the boundary and an additional "suction-off" of a diffusing substance from the boundary into the volume. By average diffusion characteristics obtained experimentally, the Fisher model makes it possible to calculate the parameters of the boundary diffusion. All diffusion experiments are carried out at sufficiently low temperatures when the contribution of boundary diffusion is predominant.

The major findings secured from the study of boundary diffusion are as follows:

1. The boundary diffusion coefficient is much larger than the bulk diffusion coefficient. The ratio D'/D in most cases varies depending on the temperature in the range of 10^3–10^5.

2. The activation energy of boundary diffusion is noticeably less than the activation energy of bulk diffusion. The ratio Q'/Q varies within 0.35–0.7 for various solvents and diffusing substances and increases slightly for solvents with a BCC lattice (0.55–0.7) as compared to that with an FCC lattice (0.35–0.55), and also rises with increasing melting point of the solvent. Some examples of the activation energy of boundary diffusion and bulk self- and hetero-diffusion are listed in Table 4.2.

Table 4.2: Activation energy of boundary and bulk diffusion.

Solvent	Diffusing substance	Q' (kcal/ g-atom)	Q (kcal/ g-atom)	Q'/Q
Ni (FCC)	Ni	26	65.8	0.40
Fe_γ (FCC)	Fe	38	67.9	0.56
Fe_γ (FCC)	Co	33	72.5	0.46
Cu (FCC)	Zn	24.5	45.6	0.54
Fe_α (BCC)	Fe	40	60	0.67
Fe_α (BCC)	Co	47	63.0	0.74
Fe_α (BCC)	Ni	43.3	61.9	0.70
Cr (BCC)	Cr	46	73.2	0.63
Mo (BCC)	W	77	112.9	0.68

The experimental data on boundary diffusion does not contradict the vacancy diffusion mechanism. However, the issue is still not completely clear.

Recently, significant progress has been made in the study of the relationship between the structure of grain boundaries and their diffusion permeability. The latter has been established to vary non-monotonically depending on the misorientation angle of the grains. Figure 4.14 represents the dependence of the activation energy Q' of zinc diffusion along the <111> slope boundaries in aluminum on the misorientation angle Θ of the grains in the temperature range of 250–340 °C (0.56–0.66 T_{melt}).

Q', kcal/g-atom

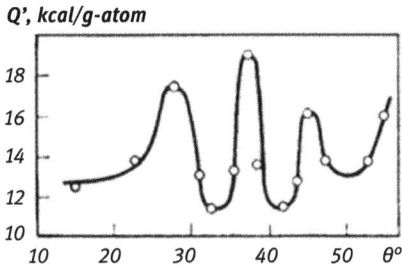

Figure 4.14: Dependence of the activation energy Q' of zinc diffusion over the < 111 > slope boundaries in Al on the misorientation angle Θ of the grains.

At angles of 28.0°, 37.5°, and 45.0°, there are clear maxima. Along electron microscopic observations, this result confirms the concept of a high-angle grain boundary as a region with a crystal structure. This concept is reflected in the model of the coincident-site lattice (for more details on this, see Chapter 5). So, the study of diffusion provides information on defects in the crystal structure, in particular, on the structure of high-angle grain boundaries.

It should be emphasized that there is a difference in the diffusion rate along immovable and moving grain boundaries. It has been found that the diffusion coefficient along the migrating boundary is 3–4 orders of magnitude greater than that along the immovable boundary. Atoms in the migrating boundary are presumed to be able to jump over long distances, which lead to the specified difference.

4.7.4 Surface

In studying diffusion along the outer surface of a metal, a high mobility of atoms of the base metal as well as of adsorbed atoms (adatoms) has been found. Surface diffusion occurs even faster than the boundary one. Figure 4.15 outlines the dependencies of ln D on 1/T for surface, boundary, and bulk self-diffusion of silver.

The activation energy of the surface self-diffusion is two times less than the boundary one, and 4.5 times less than the bulk one.

Diffusion over the surface depends on its structure. The surface of a real crystal is heterogeneous. It has steps, breaks, and adatoms. An important parameter is the number of nearest neighbors of an atom diffusing along the surface. For example, the number of nearest neighbors is three if an adatom rests in the (111) plane of an FCC metal. Such atoms are very mobile; the mechanism of their motion is called the rolling-stone mechanism. This is the case of the fastest surface diffusion.

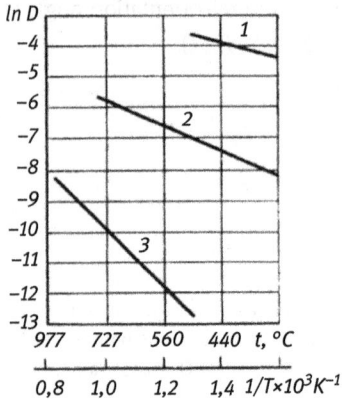

Figure 4.15: Dependence of ln D on $1/T$ for various types of silver self-diffusion: 1, surface self-diffusion, $Q = 10.3$ kcal/mol; 2, boundary self-diffusion, $Q = 20.2$; 3, bulk self-diffusion, $Q = 46$ kcal/mol.

4.8 Diffusion and equilibrium diagrams

This section covers the changes in the diffusion coefficient depending on the concentration in binary alloys and also describes some experiments showing the complexity of diffusion processes in multicomponent systems. Diffusion growth of phases is exemplified.

4.8.1 Diffusion coefficient in binary alloys

The measurement of diffusion coefficients in alloys of binary systems led to some general conclusions.

1. If adding component A to component B lowers a melting point, the diffusion coefficient increases at any given temperature (Figure 4.16).
   On the contrary, if component A raises a melting point, the diffusion coefficient of the alloy decreases. This rule is a type of observation that the diffusion coefficient at a melting point is approximately constant for a given crystal structure.
2. For a given solvent at a given temperature and composition, the diffusion process proceeds much faster in metals with a BCC lattice than in metals with a close-packed lattice. This is true for both substitutional solid solutions and for interstitial solid solutions. For example, the ratio of the self-diffusion coefficients for iron in BCC and FCC lattices at 850 °C differs by two orders of magnitude:

$$D(\alpha)/D(\gamma) \approx 10^2.$$

The ratio of carbon diffusion coefficients in BCC and FCC lattices at 910 °C is the same as the above one.

If the system has more than one phase and they are balanced, the condition of the equality of chemical potentials for all components in each phase allows one to

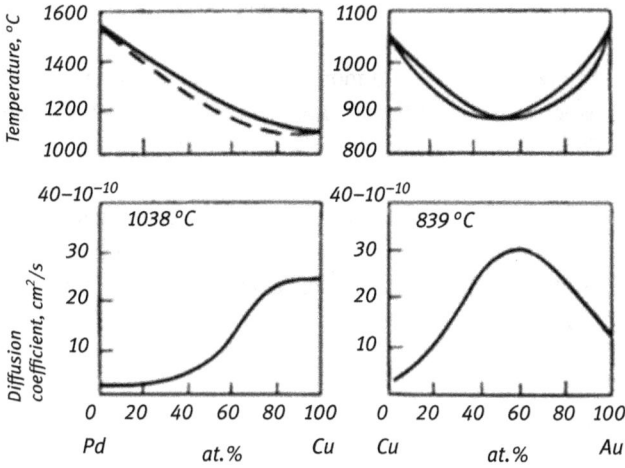

Figure 4.16: Unlimited solubility phase diagrams and the change of the diffusion coefficient depending on concentration.

determine the concentration at the boundaries of each phase. There will be a jump in the concentrations. Figure 4.17 illustrates a relationship between the phase diagrams of binary alloys and the concentration distribution curves during diffusion.

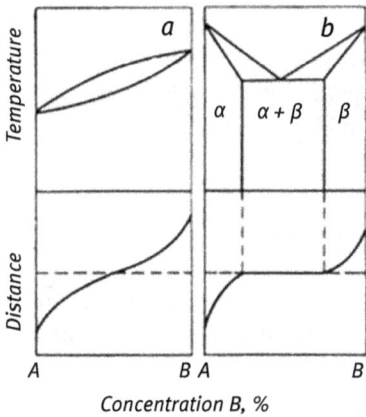

Figure 4.17: A relationship between phase diagrams and diffusion concentration distribution: (a) unlimited solubility and (b) a two-phase region $(\alpha + \beta)$.

In Figure 4.17a, a smooth curve describes the concentration distribution, and Figure 4.17b depicts a jump in the concentrations.

4.8.2 About driving forces of diffusion: Darken's experiment

Diffusion processes in three-component alloys are more complicated as compared to those in binary systems. Let us look at an instructive experiment of Darken. A sample consisting of tightly coupled Fe + 0.4% C and Fe + 0.4% C + 4% Si alloys was subjected to diffusion annealing treatment at 1,050 °C. At this temperature, both alloys are in a single-phase FCC state (austenite). A carbon concentration gradient being absent, no carbon flux can exist. This emanates from Fick's first law. However, upon annealing for 13 days, the carbon concentration gradient arose in the sample at the interface.

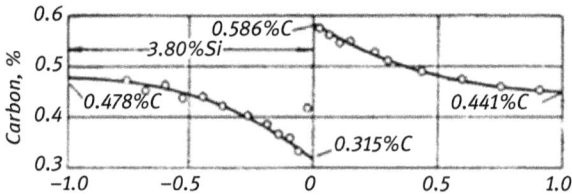

Figure 4.18: Distribution of carbon upon annealing for 13 days at 1,050 °C.

Figure 4.18 demonstrates the distribution of carbon concentration, including a sharp jump in the concentration at the interface. This effect is due to an increase in the chemical potential of carbon as silicon is added to the alloy. A local equilibrium was provided by carbon redistribution at the interface because the concentration of silicon changed dramatically. Therefore, a local change in the carbon concentration is also necessary for the local equilibrium to happen. This example shows that in analyzing diffusion, the first diffusion equation should be used as follows:

$$J = -D\,(\partial\mu/\partial x),$$

where μ is the chemical potential of the diffusing element, and D is the effective diffusion coefficient.

It would be interesting to trace the composition changes at two points A and B on opposite sides of the composite sample using a ternary diagram as schematically shown in Figure 4.19.

The points initially move along the lines of constant silicon concentration. This is because carbon diffuses much faster than silicon.

4.8.3 About diffusion growth of phases

Supersaturated solid solutions when decomposed provoke problems to be solved associated with the growth of a new phase (see Chapter 6). They can also be posted in the case of the diffusion saturation of a metal with an alloying element if the concentration

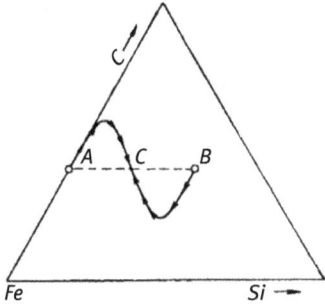

Figure 4.19: A schematic diagram of the concentration change in two points on opposite sides from the interface plane in the diffusion pair of Darken: Fe + 0.44%C and Fe + 0.48%C + 3.8%Si.

of the latter exceeds the solubility limit. In practice, new-phase growth problems are extremely important for thermal and chemical heat treatment to carry out. A distinguishing feature of the processes is to combine diffusion with phase transformations or with chemical compounds formed. During diffusion, atoms move to distances far exceeding the interatomic ones. The processes of phase or chemical transformations occur at the separation boundary.

The growth process is multistaged and includes, as a minimum, a reconstruction of the lattice at the separation boundary, as well as a diffusion supply and removal of atoms of the solute provide that the emerging and initial phases differ in composition. Usually, the lattice is thought to reconstruct rather quickly, and diffusion controls the process rate. Then, the process of the growth of new-phase particles is purely diffusional. To compute the growth rate at a moving phase interface, a substance balance equation is composed. For simplifying the problems, the shape of the boundary should be chosen flat or spherical.

Let us give examples of diffusion growth of phases.

1. During oxidation of a metal, a protective oxide layer is often formed on the metallic surface.

Figure 4.20 schematically shows the distribution of oxygen concentration in a gas (oxidizing) environment, an oxide, and a metal.

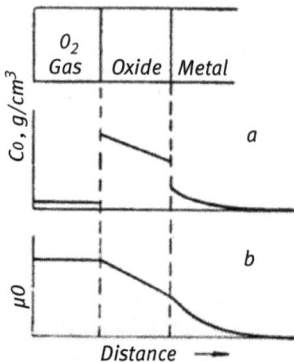

Figure 4.20: A scheme of oxygen concentration distribution in depth during oxidation of a metal (a) and the change in the oxygen chemical potential (b).

At the phase separation boundary, the concentration of oxygen changes abruptly, but the chemical potential remains constant in the contacting phases. This is because of the two-phase equilibrium rule. In iron, against the process of the formation of a protective oxide layer, the oxidation process is significantly more complicated. The oxidation of iron produces several oxides: Fe_2O_3, Fe_3O_4, and FeO. The formation of the oxides obeys the Fe–O state diagram, and the kinetics of the process is described by the parabolic law $x \sim \sqrt{Dt}$ that is usually used for approximately estimating the diffusion paths.

2. Nitrogen saturation of the surface of iron is one of the methods of chemical heat treatment, nitriding. The purpose of the latter is to increase hardness and wear resistance of materials.

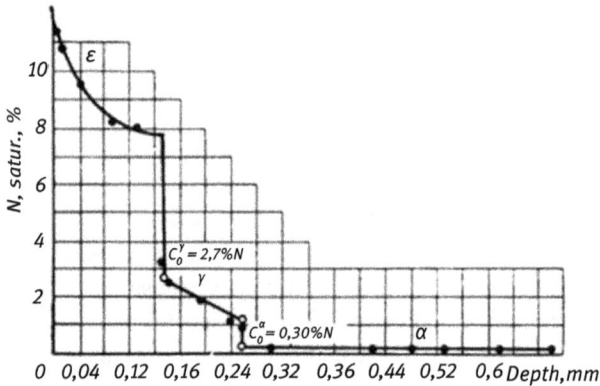

Figure 4.21: Nitrogen concentration distribution in depth of a nitrided iron layer (by Lahtin).

Figure 4.21 outlines the distribution of the concentration of nitrogen in a nitrided iron layer at a temperature of 700 °C. At this temperature, three phases are stable in the Fe–N system: (1) an ε-phase is a nitride of variable composition (~ Fe_2N) with a hexagonal lattice of metal atoms, (2) a nitrogen solid solution in FCC γ-Fe, and (3) a nitrogen solid solution in BCC α-Fe. At the boundaries of the contacting phases, jumps in the concentration of nitrogen are visible. The formation of phases in the nitrated layer obeys the Fe–N phase diagram, and diffusion of nitrogen in iron governs the nitriding kinetics.

4.8.4 The Kirkendall effect

In conclusion of this chapter, we will describe an interesting effect, which gives information about the atomic diffusion mechanism. Ernest Kirkendall was the first to design the following experiment. He took a rectangular block of brass (Cu–30%Zn) and wound it with a thin molybdenum wire (molybdenum is insoluble in copper and brass). Further, he electroplated the block with a layer of pure copper.

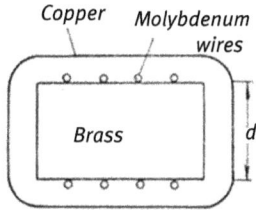

Figure 4.22: The experiment of Kirkendall and Smigelskas. Scheme of the cross section of the diffusion pair.

The sample was subjected to a series of successive anneals for different periods of time. After each annealing, a piece was cut off from the block, and the distance d between the wire marks was measured (Figure 4.22). It was turned out that the distance d decreased directly proportionally to the square root of the annealing time. The measurements showed that the copper layer increased at the expense of brass. This means that the flux of zinc atoms through the original separation surface is greater than that of copper atoms in the opposite direction. Hence, the diffusion coefficient of zinc is larger than that of copper.

Lawrence Stamper Darken performed an analysis of this experiment. He found that the chemical composition does not change in the vicinity of the mark and amounts to 22.5% Zn. For 56 days at 785 °C, each mark shifted by 0.0125 cm. L.S. Darken determined that D (Cu) $= 2.2 \times 10^{-9}$ cm^2/s, D (Zn) $= 5.1 \times 10^{-9}$ cm^2/s, that is, $D(\text{Zn})/D(\text{Cu}) = 2.3$.

The experiment, repeated many times, using different diffusion pairs and different materials as marks resulted in the same outcomes. The distance between the marks always diminished in direct proportion to the square root of the annealing time.

The Kirkendall effect excludes the exchange and cyclic diffusion mechanisms because of the equality of the diffusion component coefficients. With the vacancy mechanism, such equality is not necessary: the exchange frequencies of different types of atoms with vacancies may be different. Similarly, the diffusion coefficients of different components may vary for the interstitial mechanism.

Chapter 5
Plastic deformation and recrystallization

5.1 Elastic and plastic deformation: definitions

Deformation is any change in the shape and size of a body as a result of mechanical action. If deformation disappears after mechanical stresses stop acting, this is elastic deformation. It obeys Hooke's law. Hooke's law under tension has the form

$$\sigma = E\varepsilon,$$

where σ is the normal stress, E is the modulus of normal elasticity, and ε is the relative deformation (elongation). Hooke's law in shear is given by

$$\tau = G\gamma,$$

where τ is the shear stress, G is the shear modulus, and y is the shear angle.

If deformation and its consequences persist after the action of stresses, this is plastic deformation. For metallic materials, plastic deformation is usually accompanied by strain hardening or work hardening. The latter is due to irreversible changes in the crystal structure. In this case, lattice defects arise, mainly dislocations, as well as vacancies and interstitial atoms. It should be noted that the work hardening can be created by such plastic deformation that does not change the shape and size of a body. Examples include deformation under the action of shock waves or equal channel angular pressing. Therefore, a more general definition of deformation in mechanics is a change in the relative position of points of a body as a result of mechanical action. The simplest types of deformation are stretching, compression, shear, bending, and torsion.

The value (degree) of plastic deformation under tension can be calculated by the formula:

$$\varepsilon = (L - L_0)/L_0,$$

where ε is the relative deformation, L_0 is the length of a sample before stretching, and L is the length of the sample after stretching. This expression is valid if $\varepsilon \ll 1$. For any values of the degree of deformation, the formula appears as follows:

$$\varepsilon = \ln(L/L_0).$$

Plastic deformation is said to be cold, warm, or hot, depending on the deformation temperature. The borderline between these types of deformation is not clearly defined, and it depends on the mobility of the dislocations.

The temperature rearrangement of the dislocation structure during deformation is determined by dislocation creep. The latter, in turn, depends on the process of self-diffusion. Estimation of the temperature limit, above which the dislocation creep is

https://doi.org/10.1515/9783110758023-005

possible to observe, gives the value $T_{\text{creep}} \approx 0.5T_{\text{melt}}$, where T_{creep} is the creeping temperature and T_{melt} is the melting point. Below this temperature (it is below $0.3T_{\text{melt}}$ for technical metals and alloys), deformation is cold. At temperatures $(0.3–0.6)T_{\text{melt}}$, deformation is thought to be warm. In this case, the deformation manifests as the fine-structure arrangement though the dislocation creeps but almost without the migration of boundaries. Deformation at temperatures above $(0.6–0.7)T_{\text{melt}}$ is hot; it is accompanied not only by intensive restructuring of the dislocation structure but also by the migration of boundaries and sub-boundaries associated with the development of recrystallization.

The main mechanism of cold plastic deformation of metals is the intragrain shear displacement of some parts of a crystal (crystallite) relative to others. This, chiefly, is sliding and twinning, as well as a special type of deformation, deformation by martensitic transformation (see Chapter 6). For large degrees, in addition to shear movement, rotational deformation modes are included, which are brought about by rotating the volumes of the crystal. At high temperatures, there is also a diffusion movement of individual crystallites (grains) relative to each other. Therefore, during warm and hot deformation, the microstructure is formed by the dislocation and diffusion deformation mechanisms. Below, we will look at cold plastic deformation.

5.2 Plastic deformation of single crystals

5.2.1 Slip

Slip or shear translational movement of some parts of a crystal relative to others is the most common deformation mechanism. Slip usually occurs along close-packed planes in the close-packed direction. A slip plane and a slip direction constitute a slip system. Table 5.1 shows examples of slip systems for the most common metals with FCC, BCC, and hexagonal lattices.

Table 5.1: Slip systems in FCC, BCC, and hexagonal crystals.

Lattice type	Metal	Slip plane	Slip direction
FCC	Cu, Ni, Ag, Au, α-brass, γ-Fe	{111}	<110>
BCC	α-Fe, Mo, W, Nb, Ta, Cr, β-brass	{110} {112} {123}	<111>
Hexagonal c/a > 1.633	Zn, Cd	(0001) base slip	
Hexagonal c/a < 1.633	Mg, Be	{10-10} prismatic slip	

If slip occurs along the same system, slip traces or numerous parallel lines appear on the surface of a crystal (and in its structure) (Figure 5.1a). They can group into strips or slip bands. At certain orientations of the crystal relative to the direction of the applied stress, slip takes place over several systems, this is multiple slip (Figure 5.1b). Slip exhibits as multivariate motion of dislocations. The dislocations move in slip planes, and their Burgers vectors are parallel to the slip direction.

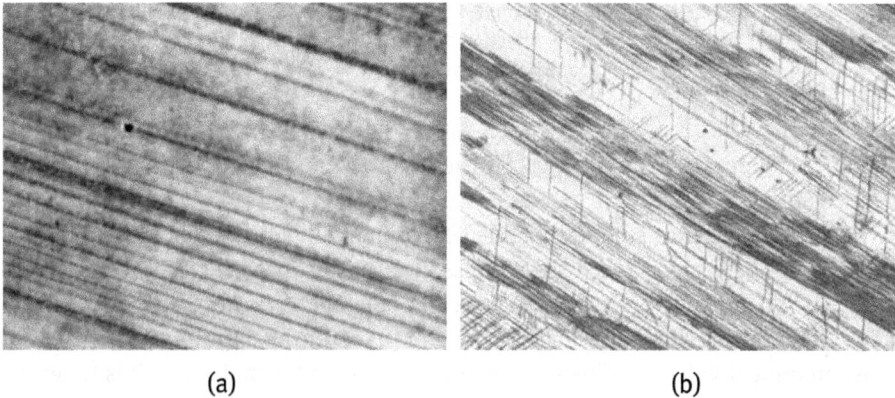

(a)   (b)

Figure 5.1: Surface of a single crystal of Al after deformation by 20% at 90 K. (a) Lines of thin slip are seen at ×12,500 magnification and (b) multiple slip in a crystal of Al; tension is along the cubic axis ($\times 100$).

5.2.2 Schmid's law

Active plastic deformation is described by a stress–strain dependence or by an $\sigma\,(\varepsilon)$ curve. Tangents drawn to this curve at different points are responsible for the hardening coefficient. The stress–strain curves substantially depend on the orientation of a single crystal. Therefore, it would be more logical to use reduced shear–stress instead of externally applied one. Figure 5.2 illustrates a diagram of the relationship between applied tensile stress and reduced shear stress.

The single crystal is oriented for easy slip. It is not hard to show that

$$\tau = \sigma \cos \phi \, \cos \psi,$$

where τ is reduced shear stress, σ is flow stress in tension, ϕ is the initial angle between the normal to the slip plane and the tensile axis, and ψ is the initial angle between the slip direction and the tension axis. This relationship is sometimes called Schmid's law. From the figure, it is seen that at $\phi = 90°$ (the tensile axis lies in the slip plane) or at $\psi = 90°$ (the tensile axis is perpendicular to the slip direction), no slip should occur. This is because the shear stress in the slip direction is equal to zero. The maximum shear stress is observed when the product is $\cos \phi \, \cos \psi = 0.5$,

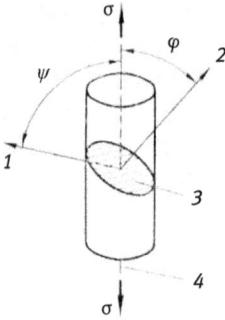

Figure 5.2: A single crystal under uniaxial tension, oriented for unit slip: 1, the direction of slip; 2, the normal to the slip plane; 3, the slip plane; and 4, the tensile axis.

which corresponds to the values of the ϕ and ψ angles equal to 45°. Hence, $\tau_{max} = 0.5\sigma$.

If one reduces flow stresses obtained experimentally for crystals of different orientations to the same reduced shear stress, it turns out that the critical shear stress is a constant for a given metal. That is, the crystals begin to plastically deform when the reduced shear stress along the slip plane in the slip direction reaches a constant critical value τ_0. This statement is also called Schmid's law. It is noteworthy that τ_0 increases dramatically with decreasing metal purity.

5.2.3 Twinning

At low temperatures and/or high strain rates, another shear deformation mechanism, twinning, is intensively developed. Twinning is especially characteristic of metals with a hexagonal and body-centered lattice.

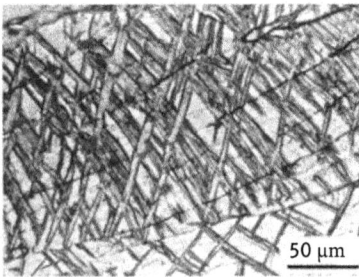

Figure 5.3: Twins in Ti, formed during high-rate deformation.

Figure 5.3 shows the microstructure of titanium with numerous deformation twins formed during high-rate deformation. Twins create "correct" geometric constructions. The boundaries of the twins are straight and parallel to each other. The location of twins in each grain indicates an orientational relationship between the crystal lattices of the matrix and the twins.

Twinning is the formation of reoriented regions in a crystal during deformation (or annealing) without changing the crystal lattice type. The reoriented regions are called deformation twins (or annealing twins); the twinning planes are the mirror reflection planes. The plane and direction of twinning constitute a twinning system. Twinning systems for a number of metals with FCC, BCC, hexagonal, and rhombohedral lattices are listed in Table 5.2.

Table 5.2: Planes and directions of twinning in metals.

Lattice type	Metal	Twinning plane	Twinning direction
FCC	Cu, Ni, Ag, Au, α-brass, γ-Fe	$\{111\}$	$\langle 112 \rangle$
BCC	α-Fe, Mo, W, Nb, Ta, Cr, β-brass	$\{112\}$	$\langle 111 \rangle$
Hexagonal	Mg, Be, Zn, Cd, Zr, Mg, Ti	$\{10\bar{1}2\}, \{10\bar{1}1\}, \{11\bar{2}2\}$	$\langle 10\bar{1}1 \rangle, \langle 10\bar{1}2 \rangle, \langle 10\bar{2}3 \rangle$
Rhombohedral	Bi	$\{110\}$	$\langle 001 \rangle$

The twinning process proceeds through the cooperative motion of atoms. Individual atoms move relative to each other with only a fraction of the interatomic distance but the total resulting shift is macroscopic. It should be noted that the atoms migrate relative to each other during slip by one or more interatomic distances.

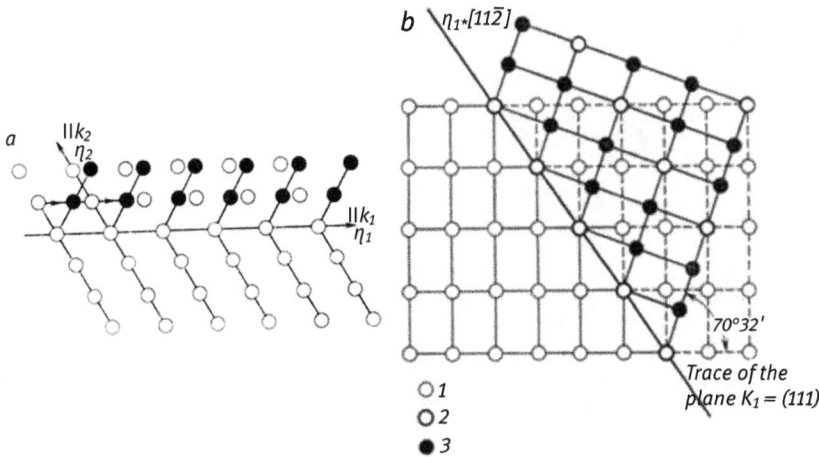

Figure 5.4: The restructuring of a crystal lattice during twinning: the bright circles are the initial positions of the atoms before twinning, the dark ones depict twinning positions of the atoms (a); the twinning geometry in the FCC lattice: 1, positions of atoms before twinning; 2, atoms are in coinciding positions; 3, twinning positions of the atoms (b).

Figure 5.4a displays a simple diagram illustrating the restructuring of a crystal lattice during twinning. The bright circles are the initial positions of the atoms; the dark ones depict the positions of atoms after twinning. The twinning plane K_1 (also known as the mirror reflection plane) perpendicular to the plane of the figure is the first undistorted plane. The twinning direction is designated by η_1. K_2 denotes the second undistorted plane. During twinning, this plane rotates by an angle, and the magnitude of which characterizes the shear deformation. Figure 5.4b shows the twinning geometry in the FCC lattice. The twinning plane is $K_1 = (111)$, the twinning direction is $\eta_1 = \{11\bar{2}\}$, and the shear angle is equal to 70°32′.

A more detailed analysis of twinning crystallography and the dislocation growth mechanism of the twin can be found in the special literature.

5.2.4 Kink bands

A slip process can be accompanied by the formation of kink bands. They are crystal regions rotated at angles that violate the laws of symmetry. Therefore, band kinking is attributed to plastic deformation by irregular rotation of the crystal lattice. The lattice in the kink bands is rotated around an axis lying in the plane of slip and perpendicular to the direction of slip. The image in Figure 5.5 involves the microstructure of a deformed aluminum crystal with kink bands.

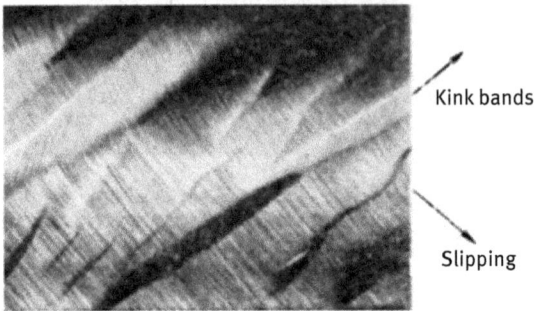

Kink bands

Slipping

Figure 5.5: Kink bands in a crystal of Al, deformed by tension to 17.5% (× 100).

The mutual arrangement of the kink bands and slip planes can be seen. The distance between the kink bands and the size of these bands depend on many factors such as the degree and temperature of deformation, the orientation of the crystal, and the degree of purity of the metal.

5.2.5 Stress–strain curves

Figure 5.6 shows the characteristic form of stress–strain curves for high-purity crystals, favorably oriented for easy slip.

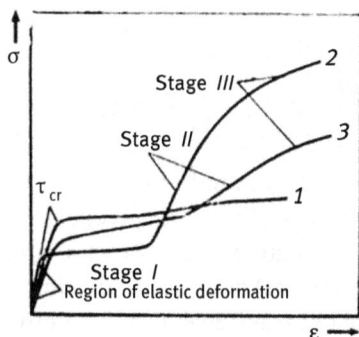

Figure 5.6: Scheme of stress–strain–deformation curves for single crystals favorably oriented for easy slip: 1, hexagonal crystals; 2, FCC crystals; 3, BCC crystals.

The reduced shear stresses τ are plotted along the ordinate axis. The $\tau(\varepsilon)$ curve for hexagonal crystals is designated as 1, 2, and 3 represent the $\tau(\varepsilon)$ curves for FCC and BCC crystals, respectively. The $\tau(\varepsilon)$ curve for FCC crystals has three distinct hardening stages. These stages belonging to the other curves are less pronounced. The type of the curves and the proportion of each stage in the total strain depend on a number of factors such as

– lattice type;
– crystal orientation;
– temperature and strain rate;
– the degree of purity, amount, and nature of impurities and alloying additives;
– stacking fault energy;
– a c/a ratio for hexagonal crystals.

At the first stage, the plot $\tau(\varepsilon)$ consists of two linear segments, but, namely, an elastic (reversible) deformation segment and a plastic (irreversible) deformation segment. The transition from the first to the second segment corresponds to a sharp bend of the curve $\tau(\varepsilon)$. In the first segment, the hardening coefficient is equal to the shear modulus G, in accordance with Hooke's law. In the second segment of the first stage, the hardening coefficient of hexagonal and FCC crystals is very small and amounts to $\approx 10^{-4}G$. Therefore, the first stage bears the name of the stage of easy slip. At this stage, the hardening coefficient for BCC crystals is greater; it is $\approx 10^{-3}G$. The critical shear stresses and the magnitude of the maximum shear strain are important characteristics of this stage. They depend on the nature and purity of a metal, on the orientation of a crystal, on the conditions of deformation, and can vary within wide limits. In the second stage, the $\tau(\varepsilon)$ dependence is also linear, but the hardening coefficient increases by about 30 times. At the third stage, the $\tau(\varepsilon)$ dependence has a parabolic shape, and the hardening coefficient diminishes. The third stage ends with the destruction of the sample.

The analysis of changes in the dislocation structure at different stages of the hardening curve is set forth in the special literature on the dislocation mechanisms of plastic deformation.

5.3 Plastic deformation of polycrystals

Stress–strain curves of polycrystals cannot be built by simply averaging the stress–strain curves of individual crystallites. Each crystallite of a polycrystal is oriented relative to applied stress differently as compared to the neighboring crystallites. Therefore, slip in different crystallites (grains) does not begin simultaneously and goes over differently space-oriented slip systems. Consequently, the resizing of different grains is not the same in different directions. Hence, each grain is exposed not only to external stresses but also to complex impacts of the neighboring grains. In order to ensure the integrity of the material and the deformation continuity during the transition through the grain boundaries, multiple slip should begin at the initial stages of the deformation, starting near the boundaries. Therefore, there is no stage of easy slip in polycrystals.

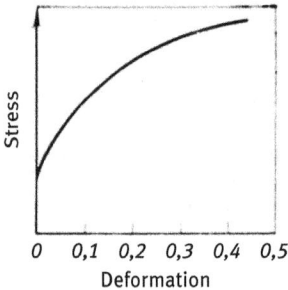

Figure 5.7: Stress–strain curve for polycrystalline FCC- and hexagonal lattice metals.

Figure 5.7 represents the characteristic stress–strain curve for FCC and hexagonal polycrystals. As can be seen, the first stage is lacking but the second and third stages take place. With cold plastic deformation, the grain boundaries are sturdier than the internal volumes of the grain. The strength of the boundaries depends on the misorientation angle of neighboring grains, as well as on the segregation of impurities in real alloys. The grain boundaries serve as places for the braking of dislocations and multiple slip, and therefore the dislocation density (and hardening) near the boundaries is higher than in the volume of the grains.

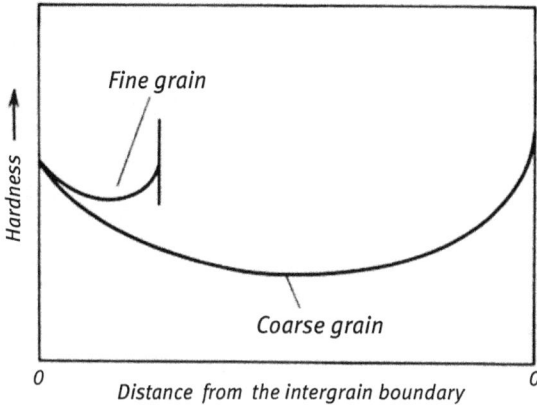

Figure 5.8: Size-grain influence on the microinhomogeneity of the grain deformation.

Figure 5.8 depicts the scheme for changing in hardness across the grain in a coarse-grained and fine-grained material. As can be observed in Figure 5.8, the heterogeneity of deformation over the grain increases with increasing the grain size. The central region of the coarse grain is softer than that of the fine one. Therefore, the strain hardening of the fine-grained material is larger than that of the coarse-grained one.

A sufficiently large degree of deformation inside a coarse-deformed grain causes a complex inhomogeneous microstructure to form. Apart from the heterogeneity described above, stripes with different internal structures arise along the boundaries and throughout the volume of the grains. One of the possible schemes for the structure of a grain deformed by rolling is represented in Figure 5.9. The figure outlines a cross section passing through the rolling direction and a normal to the rolling plane. The grain microstructure consists of stripes of various origins. At an angle of 30–40° to the rolling direction, shear bands emerge. They represent regions of strong shear strain localization. The shear bands intersect the grain boundaries and can pass through the entire thickness of the sheet. With an increase in the deformation degree, their number rises until they become a predominant structural component. Other bands (deformational and transitional) are located along the rolling direction (Figure 5.9).

Inside the deformation bands, a cellular structure is formed. Inside the shear band and transition bands, microbands are observed. Neighboring microbands in the transition band are misoriented at angles up to 5°. However, the turn between adjacent microbands has the same sign; the misorientation accumulates when passing from one edge of the transition band to another, and can reach several tens of degrees. The causes and mechanism of the formation of the microbands have not been reliably established.

Describing the behavior of metals during plastic deformation is too involved. This is because there is no transition from dislocation deformation physics to macroscopic

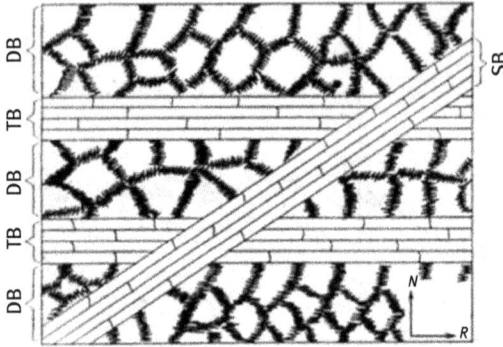

Figure 5.9: Scheme of the structure of a cold-rolled metal grain with deformational bands (DB), transitional bands (TB), and a shear band (SB): R, the rolling direction; N, normal to the rolling plane.

characteristics at the engineering level. Physical mesomechanics developed in the last decade by Panin and his followers can propose such a transition. Physical mesomechanics is believed to organically combine the physics of plastic deformation (a microlevel), continuum mechanics (a macrolevel), and physical materials science. A multilevel approach to the description of plastic deformation underlies the fundamental principle of physical mesomechanics. This approach suggests the existence of a hierarchy of interrelated different-scale deformation levels inside a deformable body. Within this technique, the deformation inside individual grains (a microlevel) continues with collective effects covering groups of grains (a mesolevel), and further with effects covering the entire sample (a macrolevel). Another principle of physical mesomechanics includes the wavy nature of plastic deformation, according to which the deformation propagates through a sample in the form of alternating regions having different degrees of deformation.

The conceptual issues of physical mesomechanics of materials are described in detail by V.E. Panin and his disciples in the special literature.

5.4 Large plastic deformations

Recently, due to the tendency to strive for creating nanocrystalline and submicrocrystalline structures, studies of large plastic deformations have been developed. V.V. Rybin was the first to publish a monograph on large plastic deformations of metals. In the last decade, R.Z. Valiev et al. have performed active investigates of large plastic deformations. He called this deformation intensive. A.M. Glezer proposed using a different term called megaplastic deformation.

Large plastic deformations, in addition to sliding and twinning, cause rotational modes of plastic deformation. Rotational mode carriers are partial disclinations. According to V.V. Rybin, the "fundamental peculiarity of the developed plastic deformation consists neither in increasing the density of dislocations and nor in enhancing the spatial heterogeneity of their distribution but in the appearance of misorientations of crystalline microregions." Misorientations are accumulated due to the turns of these microregions relative to each other. In fact, rotational modes of plastic deformation, kink bands, and shear bands (described above) are specific examples of the action of the physical mesomechanics laws.

Structural changes such as deformation–origin boundaries and misorientations at these boundaries under large plastic deformations can be confirmed through electron microscopic observation.

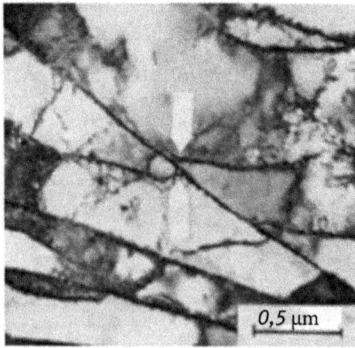

Figure 5.10: Knife boundary in deformed molybdenum ($\varepsilon = 0.9$). Misorientation at the boundary is 28.3°.

Figure 5.10 presents an example of a fragmented structure with such a boundary in molybdenum deformed to a large extent. Along with the formation of low-angle boundaries, a straight-line ("knife") high-misorientation boundary arises. The latter has a shift of a small-angle boundary (indicated by arrows). This fact evidences that the plastic turns of the mating parts of the crystal and the plastic shifts take place simultaneously. V.V. Rybin made a conclusion that "the emergence of the deformation-origin boundaries. . . should be understood as a structural response of the crystal to the development of rotational plasticity modes in it."

Intensive plastic deformation causes a cellular dislocation structure to form. This structure takes up an increasing volume of the sample as the degree of deformation increases. The size of the cells depends on many factors; it varies from hundreds to tens of nanometers and diminishes with increasing degree of deformation. Usually, the cells have blurry, "fluffy" borders. The misorientation of neighboring cells reaches several tens of degrees and increases with increasing degree of deformation.

Figure 5.11 represents an electron microscopic image of the cellular dislocation structure in aluminum. Dislocations are concentrated mainly within the boundaries of the cells. The internal cell volume is almost free of dislocations. Estimates show that the cellular structure, when formed, reduces the elastic energy of the system as compared with the chaotic distribution of dislocations, provided that the number of dislocations is the same in both cases. There is an upper limit for the size of the

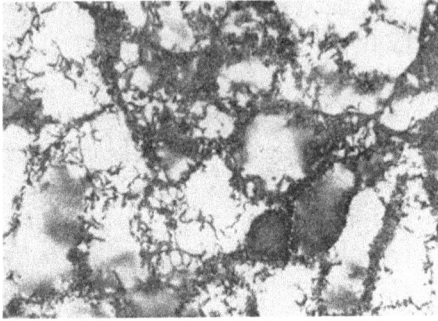

Figure 5.11: An electron microscopic image of the cellular dislocation structure in aluminum.

dislocation cells at a given degree of deformation. Typically, the cell size is an order of magnitude larger than the average distance between the dislocations in the case of their chaotic arrangement.

A diagram of the change in the structure of the cell with an increase in the degree of deformation (according to Valiev and Aleksandrov) can be viewed in Figure 5.12.

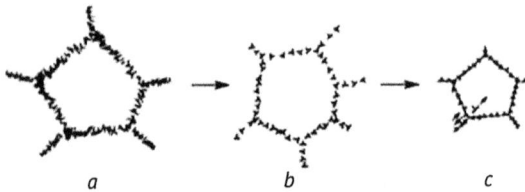

a   b   c

Figure 5.12: Scheme of changing in the structure of the cell during intensive plastic deformation.

The cellular structure formed under large deformations is transformed during intensive plastic deformation as sketched in Figure 5.12a. The walls of the cells become narrower and more ordered (Figure 5.12b). Subsequently, both the cell size and the wall thickness decrease (Figure 5.12c), the dislocation density in the walls increases, and annihilation of dislocations of the opposite sign occurs. As a result, the cell walls preserve redundant dislocations of two signs (Figure 5.12c). These dislocations play a different role. With further deformation, dislocations with the Burgers vector perpendicular to the boundary give rise to a growing number of misorientations and to the transformation of cells into grains. Slipping dislocations are responsible for long-range stress fields. The grain boundaries are in a nonequilibrium (excited) state. Around the boundaries, there are regions of strong distortions of the crystal lattice, caused by grain boundary dislocations. The slipping grain-boundary dislocations provide grain-boundary sliding and relative displacement of the grains.

A schematic representation of a nanostructured material for two different grain sizes is shown in Figure 5.13.

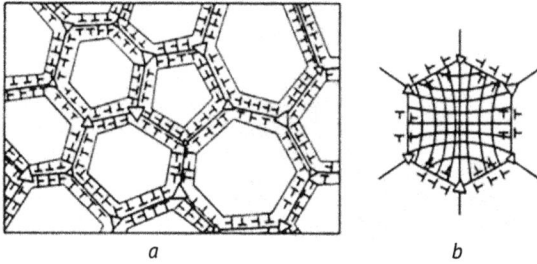

Figure 5.13: Scheme of the structure of a nanostructured material with a grain size of about 100 nm (a) and of 10–20 nm (b).

With a grain size of about 100 nm, the central part of the grain has a perfect crystal lattice. Near the boundaries, elastically distorted zones with a width of several nanometers are located (Figure 5.13a). With a grain size of 10–20 nm, the elastic distortions spread over the entire volume of the grain (Figure 5.13b). Triangles of different sizes and orientations denote disclinations of different magnitude and sign.

Intensive plastic deformation is usually carried out under high pressure to prevent the destruction of materials. The main deformation methods are equal-channel angular pressing and torsion under high pressure in Bridgman anvils. Usually, upon pressing, the grains produced are of larger size than under torsion. However, in the latter case, the deformed samples obtained have a significantly bigger size. In pure metals, equal-channel angular pressing creates a structure with a grain size from 100 to 400 nm, which is called submicrocrystalline. This technical term implies that the grain size amounts to less than 100 nm and, therefore, the nanocrystalline state is said to not be reached. An example of such a structure obtained in industrial titanium is illustrated in Figure 5.14.

Nanocrystalline and/or submicrocrystalline structure materials have high mechanical properties and are widely applied in practice. Of particular interest is the possibility of influencing the structurally insensitive properties (elastic moduli, Debye temperature, etc.).

5.5 The formation of point defects during deformation

Plastic deformation is accompanied by the formation of point defects such as vacancies and interstitial atoms, with their concentrations being several orders of magnitude higher than an equilibrium one. The mobility of interstitial atoms in metals is high, and their sinking to dislocations and boundaries takes place at low temperatures. Vacancies are significantly more stable defects. Studies on the recovery of electrical resistivity of gold after deformation at cryogenic temperature have showed that annihilation of interstitial atoms occurs already at 77 and 90 K, while

Figure 5.14: Submicrocystalline structure in industrial titanium.

annihilation of vacancies proceeds at 300 K. In commercially pure metals and alloys, the thermal stability of vacancies and their concentration depend on the presence and type of impurities. The resulting low-mobility impurity–vacancy complexes contribute to maintaining a high concentration of vacancies to higher temperatures.

The mechanisms of forming point defects during deformation have been little examined due to experimental difficulties. Several mechanisms have been proposed, which are premised on analyzing the geometry of interaction of dislocations. Firstly, point defects can form by the nonconservative movement of dislocation steps. The Burgers vector of the step has a different orientation than the vector of basic dislocations. Therefore, when moving, the step generates a series of either interstitial atoms or vacancies. This mechanism can be observed in low stacking-fault energy metals when dislocations are split into partial ones. Another mechanism of forming the vacancies is the interaction (annihilation) of two edge dislocations of opposite sign. If the dislocations move in the same slip plane, a series of interstitial atoms emerge during annihilation. If the slip planes are located by two or more interatomic distances, the annihilation results in creating a line of vacancies. A similar situation may be encountered when the edge dislocation is bent around a screw dislocation.

Point defects are responsible for the behavior of some physical properties of metals subjected to plastic deformation. They also play an important role in the processes that occur during the heating of deformed metals.

5.6 Deformation textures

A dominant crystallographic orientation of grains in a polycrystal is referred to as the crystallographic texture. Textures are characterized by ideal orientations, and the dispersion of orientations around an ideal grain arrangement characterizes the perfection of the texture. The more perfect the texture, the less its dispersion is. To describe the texture, straight pole figures, reverse pole figures, and orientation distribution functions (ODFs) are used. The straight pole figure shows how the orientation of the crystal planes of a certain type is distributed relative to the characteristic direction of the sample, but, namely, to the axis of the sample or normal to its plane. The reverse pole figure is responsible for the distribution of the orientation of certain directions in the sample relative to the crystal lattice. To construct pole figures using X-ray methods, reflective planes with low indices (100), (110), (111), and so on are usually utilized. For a more complete representation of the textures, ODFs are applied to be reconstructed from pole figures.

Textures can arise during the processes of crystallization, deformation, recrystallization, and phase transformation. When a polycrystal is deformed, its grains acquire the same shape as the sample as a whole (the Polanyi principle). The grain orientation changes in such a way that the shear direction is turned toward the action of the tensile stresses. During compression, the shear direction approaches the plane of compression. The orientation change occurs gradually as the degree of deformation rises. With a sufficiently large degree of deformation, a stable grain orientation appears; it is characteristic of this texture. The grain turns that provoke the texture to form are heterogeneous for different grains and even within a single grain. Such inhomogeneities cause dispersion of the texture.

The type of texture depends on the conditions of deformation, primarily on the type of deformation such as tension, compression, drawing, and rolling, and on the properties of a material, mainly on its crystalline structure. The most common deformation textures are the texture of drawing (an axial one) and the texture of rolling (a limited one). The former is characterized by a crystallographic direction, usually with low indices, parallel to the axis. Around the texture axis, the grains occupy an arbitrary position. So, when drawing BCC-lattice metals, the grains are turned so that the <110> direction is aligned along the drawing axis. In FCC-lattice metals, two groups of grains with different textures arise, and a <111> + <100> two-component texture is formed. The ratio of the components depends on the properties of a metal. The <111> texture predominates in high stacking-fault energy metals (aluminum). The <100> texture becomes stronger in metals with a relatively low stacking-fault energy (silver and brass). In metals with a hexagonal lattice, the basic planes as slip planes are turned to locate parallel to the drawing axis. If the slip occurs along prismatic or pyramidal planes, or if the deformation is carried out by means of twinning, the texture is more complex.

The rolling texture is characterized by a plane parallel to the rolling plane and a direction parallel to the rolling direction. In a rolled metal, several main textures and a small number of small components can coexist. BCC-lattice metals have a {001} <110> main rolling texture with a {001} plane parallel to the rolling plane and a <110> direction parallel to the rolling direction. The <110> rolling direction is the same as the axis direction of the axial texture of the BCC metals. The formation of the rolling plane is determined by slip crystallography. The textures such as {112} <110> and {111} <112> accompany the main texture.

FCC-lattice metals and alloys have the {110} <112> main rolling texture ("brass texture"). Such a texture can be observed in low stacking-fault energy alloys during difficult transverse slip (α-brass, silver, FCC-metal-based solid solutions). As cross-slip is facilitated, the {112} <111> orientation appears and enhances. In FCC high stacking-fault energy metals (copper and aluminum), a {112} <111> + {110} <112> two-component texture ("copper texture") is produced. According to M.A. Stremel, during rolling of low stacking-fault energy alloys, twinning in the shear bands takes place, but during rolling of high stacking-fault energy alloys, slip occurs. Therefore, twinning in the shear bands formed leads to a "brass texture," and slip causes a "copper texture." Contributing to twinning, lowering the temperature of deformation can trigger a transition from the "copper texture" to the "brass texture."

The main component of the rolling texture of hexagonal metals is the {0001} <1120> orientation. When rolling, the basic plane (slip plane) tends to locate parallel to the rolling plane, and the close-packed direction <1120> moves to line up parallel to the rolling direction. This was to be expected since the compressive stresses acting perpendicular to the rolling plane tend to turn the slip plane toward the sheet plane. This texture can be revealed in metals with a ratio of c/a axes close to an ideal ratio (magnesium and cobalt). Change in the c/a ratio complicates the texture.

The study of textures plays an important role in analyzing anisotropy of physical and mechanical properties, which occurs during plastic deformation. The mechanical properties in the longitudinal direction (e.g., in the rolling direction) can be much greater than in the transverse direction. A detailed description of the crystallographic textures in metallic materials is available in the book by Gunter Wassermann and Johanna Greven.

5.7 The influence of plastic deformation on the properties of metals

Under deformation of metals and alloys, the work expended turns into latent work-hardening energy and heat. The latent work-hardening energy gives rise to irreversible changes in the crystal structure (vacancies and mainly dislocations). When the deformation is insignificant, almost all the work expended is converted into latent energy. However, as the work done increases, as the degree of deformation increases, the proportion of energy absorbed decreases. This fact is associated with the interaction and annihilation of defects upon raising their density. Each type of crystal structure defects corresponds to a characteristic value of excess energy. Its value is measured based on the study of annealing of defects when the deformed metal is heated. Defects of various types are annealed at different temperatures, and this makes it possible to determine the energy value to form a vacancy, dislocation, and stacking fault. To conduct the measurements, the methods of calorimetry and electrical resistance are utilized. Such measurements are very painstaking and require great accuracy. The results depend on the presence of difficult-to-control impurities, on the temperature of deformation, and therefore are not always reproducible.

Mechanical properties change most significantly during plastic deformation. Strength characteristics such as yield strength, temporary resistance, and hardness increase several times. FCC-lattice alloys harden more strongly than BCC-lattice alloys. So, the yield strength of FCC metals may increase sevenfold. For example, the strength of σ_B during deformation increases by 90% from 420 to 1,080 MPa in 08 steel (BCC), and from 620 to 1,830 MPa in 10X18H8 steel (FCC). The degree of hardening also depends on the magnitude of the stacking fault energy. The ductility characteristics (elongation and lateral compression) are greatly reduced.

A decrease in the grain size leads to a decrease in the mean free path of dislocations, and the strength characteristics rise. An increase in yield strength can be deduced from the Hall–Petch law:

$$\sigma_T = \sigma_0 + kd^{-1/2},$$

where σ_T is the yield strength, σ_0 is the friction stress, k is the constant associated with the propagation of deformation across the grain boundaries, and d is the size of the grains or subgrains. The exponent is (−1/2) for BCC-lattice metals but may differ from (−1/2) for FCC-lattice metals and FCC-lattice alloys. Friction is created by clusters of dislocations at the grain boundaries and impurity atoms in the slip planes, by Peierls forces.

Of great interest is the enhancement of mechanical properties due to intensive plastic deformation leading to a nano- and submicrocrystalline structure. Tensile tests of nanocrystalline materials show that a significant increase in their strength is not accompanied by a strong decrease in plasticity. This is because a new channel of plastic deformation, namely, grain-boundary sliding, emerges. Nanostructured commercially pure titanium has strength properties such as temporary resistance of 1,100 MPa, and plasticity (elongation of 10%) is the same as the widely used Ti–6% Al–4%V titanium alloy after standard processing. Once equal-channel angular pressed, 12X18H10T stainless austenitic steel with a grain size less than 100 nm has yield strength of 1,300 MPa. For comparison, upon heat treatment, this steel has the yield strength of 6 times less. With such strength, this steel saves high-level plasticity (elongation of 27%). Fundamental, usually structurally insensitive physical properties may be changed in metals with a nanocrystalline structure. So, a decrease in the Curie temperature and saturation magnetization is observed in nickel, and a decrease in Young's modulus is revealed in copper.

The mechanical properties of a deformed metal are different in different directions. The anisotropy of the properties is due to both the crystallographic texture and the shape of the grains elongated along the rolling or drawing direction. The number of grain boundaries in the transverse direction is greater than in the longitudinal one. The boundaries concentrate impurities and nonmetallic inclusions that weaken the resistance to deformation. Therefore, against longitudinal specimens, transverse ones can have much lesser characteristics of plasticity and impact strength.

Work hardening causes an increase in electrical resistance due to scattering of conduction electrons by defects of the crystal structure. The measurements of electrical resistance at cryogenic temperatures, for example, at the temperature of liquid helium, when the scattering of electrons by phonons is minimized are the most illustrative. In this case, the contribution of defects and impurities to the increase in electrical resistance should be taken into account.

When deformed, crystal lattice defects increase the specific volume or decrease the density. For the purpose of obtaining data on the change in density of a deformed metal, the density is measured before and after annealing of defects. In doing so, the methods of hydrostatic weighing or dilatometry are employed. The effect of increasing the specific volume during deformation amounts to a few tenths of a percent or even less.

Each crystallite is anisotropic; its properties depend on the crystallographic direction. In a metal with a chaotic orientation of crystals, properties in all directions are statistically averaged. Such a metal is quasi-isotropic. In a textured metal, there are directions along which some properties are increased, and others are reduced. Therefore, the texture is responsible for the anisotropy of properties. Not only mechanical but also magnetic properties are anisotropic. The shapes of the hysteresis loop in the longitudinal and transverse directions are different. Typically, plastic deformation leads to an increase in coercive force, and a decrease in residual induction and magnetic permeability. This is because of delaying in the displacement of domain boundaries by defects during magnetization reversal.

5.8 Phenomena in deformed metals during heating

When a deformed metal is heated, annihilation and restructuring of defects occur, the fields of elastic stresses decrease and disappear. As a result, the latent energy of work hardening is released, and the metal passes into a more balanced state. As the temperature grows, elementary physical processes proceed in the deformed metal. They are listed below in the order of their development:

1. Diffusion of point defects and their sink into dislocations and boundaries
2. Redistribution of dislocations through simple and transverse slip
3. Redistribution of dislocations through creep
4. Polygonization or the formation of low-angle boundaries related to the processes specified in items 2 and 3
5. Primary recrystallization or migration of small-angle and intergrain high-angle boundaries to a deformed matrix, along with "sweeping" of defects
6. Collective and secondary recrystallization or migration of intergrain boundaries between recrystallized grains and the enlargement of the latter

These elementary processes can take place simultaneously. This depends on many factors such as the degree of previous plastic deformation, the heating rate and the holding time during annealing, and the purity of the material. All the processes are thermally activated. The Arrhenius equation provides their temperature-dependent rate v:

$$v = v_0 \times \exp\left(-Q/kT\right),$$

where Q is the activation energy of a given process. When joining several processes, Q characterizes the total effective activation energy.

The most important process that changes the structure of a deformed metal is primary recrystallization. In this case, deformed grains containing an increased density of defects are replaced by new, more perfect grains, as a rule, by equiaxial ones. In metallography, all processes occurring during heating of deformed metals are divided into two large groups:
1. Recovery including relaxation and polygonization
2. Primary, collective (uniform grain growth), and secondary (uneven grain growth) recrystallization

5.9 Recovery: relaxation and polygonization

Recovery is a multistage process that includes all the elementary physical processes before recrystallization. Recovery may be considered as a phenomenon of lowering the hardness and strength of a deformed metal. Recovery is related to increasing structural perfection of the metal by redistributing and reducing the concentration of point defects, as well as redistributing and partially annihilating dislocations without creating new boundaries (relaxation) or with forming and migrating only low-angle boundaries (polygonization).

Relaxation is not accompanied by changes in the microstructure, which can be detected by means of metallography and electron microscopy methods. Therefore, to understand the relaxation kinetics, it is mandatory to ascertain changes in the physical and mechanical properties of a deformed metal during annealing depending on the temperature and holding time. The measurement of electrical resistance may be served as an informative method for studying the annealing kinetics of point defects. During annealing of defects in copper and gold, deformed at temperature of liquid helium, the relaxation has experimentally been shown to be a multistage process. The stages were separated on the temperature interval and activation energy. The activation energy is the thermal energy required for removal structural defects of a given type. The activation energy Q is included in an expression for a time τ that dictates the beginning of the process depending on the temperature:

$$\tau = \tau_0 \times \exp\left(Q/kT\right),$$

where τ_0 is a constant and k is the Boltzmann constant. It should be pointed out that the relaxation has no incubation period; it begins with a maximum rate and decays with time under isothermal conditions. The following stages are fixed. At the lowest temperature (30–40 K), recombination of pairs such as interstitial atom and vacancy takes place. At temperatures of 90–200 K, an interstitial atom and bivacancies migrate; the activation energy is 0.2–0.5 eV. At 210–320 K, dislocations are redistributed without forming new boundaries, and the activation energy amounts to 0.7 eV. These stages constitute the process of relaxation. At temperatures above room temperature (350–400 K), the dislocations redistribute, partially annihilate, and form low-angle boundaries (polygonization); and the activation energy is 1.2 eV.

Interstitial atoms are more mobile than vacancies. Therefore, when relaxing, they disappear earlier than vacancies. The most important structural change during relaxing is to reduce the excess concentration of vacancies. The sink rate of excess vacancies depends on the activation energy of diffusion. Suppose that the activation energy E of diffusion linearly depends on the melting temperature T_{melt}:

$$E = \alpha T_{\text{melt}}.$$

The diffusion coefficient D has the form

$$D = D_0 \times \exp\left(-E/RT\right) = D_0 \times \exp\left(-\alpha T_{\text{melt}}/RT\right).$$

The values of D_0, α, and R are constant, and the sink rate of vacancies is determined by the ratio T_{melt}/T. The lower the melting temperature is, the greater the sink rate of vacancies at a given temperature. It follows that, in low melting-point metals (e.g., aluminum), excess vacancies disappear at room temperature. In rolled copper and nickel, excess vacancies upon deformed disappear at 20 and 100 °C, respectively.

Polygonization is accompanied by changes in the microstructure, which are recorded by electron microscopy method and, in some cases, by metallography and X-ray diffraction methods. Polygonization is the redistribution dislocations by sliding and creeping. In this case, polygonization reduces the density of dislocations and forms subgrains surrounded by low-angle boundaries. Enlargement of subgrains through migration of sub-boundaries or coalescence of a group of neighboring subgrains are also attributed to polygonization. Polygonization can be both stabilizing one and pre-recrystallization (by Gorelik). The meaning of these terms will be clear below. A diagram that illustrates stabilizing polygonization is shown in Figure 5.15.

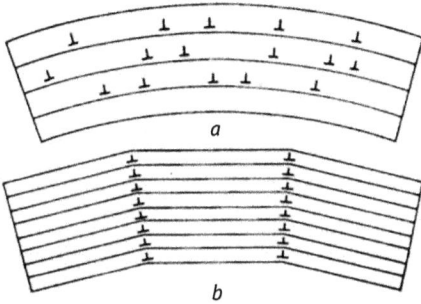

a

b

Figure 5.15: Scheme of the chaotic arrangement of edge dislocations in a crystal deformed by bending before polygonization (a) and the alignment of dislocations in the wall during polygonization (b).

An excess of edge dislocations of one sign creates a crystal bend (Figure 5.15a). During annealing, the dislocations line up one below the other in the walls (Figure 5.15b). Simultaneously, one dislocation produces the rarefaction region; under the latter, there is a condensation region formed by the neighboring dislocation (Figure 5.16).

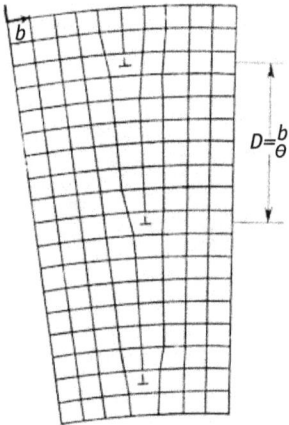

$$D=\frac{b}{\theta}$$

Figure 5.16: Structure of the low-angle boundary (sub-boundary of slope).

The fields of elastic dislocation stresses are largely compensated. The wall of dislocations does not have a long-range stress field. Consequently, the formation of dislocation walls is an energetically advantageous process. This process would seem to be proceeding spontaneously. However, its development requires thermal activation. The walls are formed as a result of a combination of slip and creep of dislocations. Creep is the slowest process that involves diffusional displacement of atoms between slip planes.

The wall of dislocations of one sign is a low-angle boundary and separates low-misorientation subgrains. As can be viewed from Figure 5.16, the misorientation angle θ and the distance D between dislocations in the wall are related as follows:

$$\theta \approx b/D,$$

where b is the Burgers vector of a dislocation. As the angle θ increases, the distance between dislocations diminishes until the boundary becomes high angle. In this case, the boundary can no longer be described through dislocations for some value of θ.

The described low-angle boundaries are the boundaries of the slope. A low-angle torsion boundary forms when emerging a series of screw dislocations instead of edge dislocations. Obviously, a set of misorientations can be obtained by combining grids of edge and screw dislocation types. In all these cases, the degree of plastic deformation is small, and a few numbers of slip planes operate.

With stabilizing polygonization, small-angle boundaries emerge; however, there were no boundaries or their contours in the deformed state. A characteristic feature of the sub-boundaries that arise is their small curvature and, therefore, their low mobility. The resulting structural state is stable and inhibits recrystallization as temperature further rises. Such polygonization competes with recrystallization and therefore is referred to as the stabilizing one.

After medium and large degrees of deformation, when multiple slip is realized and a cellular structure arises, the polygonization mechanism is different (Figure 5.17).

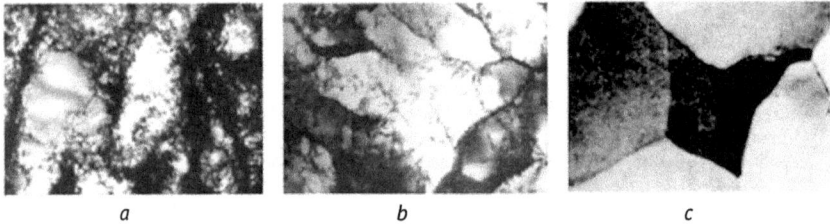

$$a \qquad\qquad\qquad\qquad b \qquad\qquad\qquad\qquad c$$

Figure 5.17: Pre-recrystallization polygonization. Transformation of cells (a) into subgrains (b, c) ($\times 2,500$).

The contours of the future subgrains are largely already formed by the previous deformation. When heated, the bulk dislocation plexuses surrounding the cells flatten and turn into sub-boundaries, and the cells turn into subgrains (by Gorelik). The sub-boundaries that arise have a large curvature and relatively high mobility. If such subgrains are poorly misoriented, they can be enlarged either according to the mechanism of accretion of neighboring boundaries or due to coalescence. Many subgrains serve as recrystallization centers during subsequent heating, and therefore this polygonization is called pre-recrystallization.

In conclusion of this section, we would like to note that impurities significantly affect the temperature of polygonization. So, polygonization begins at 200 and 850 °C in high-purity iron (0.001% C) obtained by zone melting and in Armco iron (0.05% C), respectively.

5.10 Recrystallization

Recrystallization is a process of replacing some grains of a given phase by more perfect in structure and lower energy grains of the same phase. The recrystallization process is accomplished by the appearance and movement of high-angle misorientation boundaries (primary recrystallization) or only the movement of such boundaries (collective and secondary recrystallization).

The recrystallization rate is determined by the gain of the free energy ΔF. This gain plays the role of a driving force. Let us denote the quantity ΔF referred to 1 mol through P_{df}. P_{df} consists of several components:

$$P_{df} = \sum P_i = P_{blk} + P_{gr} + P_{surf},$$

where P_{blk} is the gain in bulk energy, P_{gr} is the gain in grain boundary energy, and P_{surf} is the gain in surface energy. The meaning of these terms will be clarified below.

Let M be the mobility of the boundary, affected by the structure of the boundary. P_{br} is a force that impedes the migration of the boundary, primarily due to dissolved impurities and particles of excess phases. The velocity V of the boundaries, and hence the recrystallization rate can be written as follows:

$$V = M \sum P_i.$$

For primary recrystallization

$$V_{prim} = MP_{blk};$$

for collective recrystallization

$$V_{coll} = MP_{gr};$$

for secondary recrystallization

$$V_{sec} = MP_{gr} \ \text{or} \ V_{sec} = MP_{blk} \ \text{or} \ V_{sec} = MP_{surf}.$$

Making an allowance for the braking forces, the boundary velocity acquires the final form:

$$V = M(P_{df} - P_{br}).$$

5.11 Primary recrystallization

Primary recrystallization is a process of forming and growing new, distortion-free, and much more perfect grains in a deformed matrix, with the grains separating from the matrix by high-angle misorientation boundaries. During plastic deformation, stress fields arise around dislocations, and excess free energy is accumulated and localized in them. The excess free energy is the driving force of the recrystallization process. Therefore, the driving force of primary recrystallization P_{df}^{prim} can be written as follows:

$$P_{df}^{prim} \approx k\Delta N_{\partial} G \, b^2,$$

where ΔN_{∂} is the dislocation density difference in the deformed matrix and the recrystallization nucleus, G is the shear modulus, b is the Burgers vector, and k is the coefficient depending on the dislocation distribution, $0.1 \leq k \leq 1$. The dislocation density in a deformed metal amounts to 10^{11}–10^{12} cm^{-2}. After recrystallization, it drops to 10^6–10^8 cm^{-2}.

A large amount of experimental data on primary recrystallization can be summarized in the form of the following seven statements:

1. There is minimum deformation provoking recrystallization.
2. The smaller the degree of deformation is, the higher the temperature needed for the onset of recrystallization.
3. With annealing time prolonged, the onset of recrystallization requires a lower temperature.
4. After the completion of primary crystallization, the final grain size is affected by mainly the degree of deformation and annealing temperature to a lesser extent. The grain size is the smaller, the higher the degree of deformation and the lower the annealing temperature. This fact is explained by the following pattern. With an increase in the work hardening rate, the nucleation rate of new grains increases faster than their growth rate. However, as the annealing temperature further rises, the growth rate accelerates more dramatically.
5. The larger the grain size before deformation, the higher the degree of deformation required for recrystallization.
6. The higher the deformation temperature, the more difficult it is to cause recrystallization upon subsequent heating.
7. The newly formed grains cannot grow inward the deformed grains with the same orientation or inward the grains with an orientation close to the twin one. This is because of the low mobility of low-angle and twin boundaries (see below).

Under isothermal holding, the recrystallization kinetics is described by the dependence shown in Figure 5.18a. For comparison, Figure 5.18b sketches an analogous dependence for relaxation.

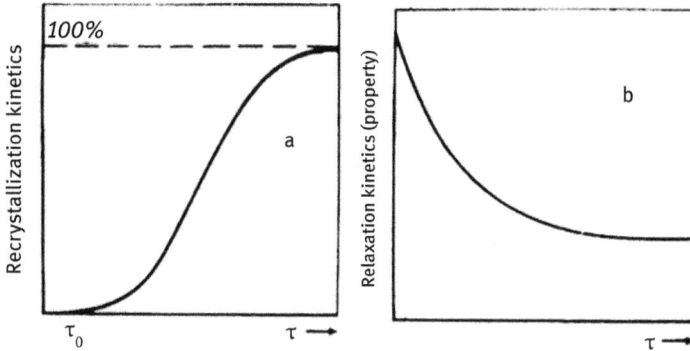

Figure 5.18: Time dependencies of the degree of recrystallization (a) and relaxation (b) during isothermal holding.

Against relaxation, recrystallization has an incubation period, then accelerates, and then slows down. The incubation period is associated with the formation of recrystallization nuclei. The acceleration is due to the intensive growth of new grains from the nuclei arisen. The slowdown occurs when full-grown grains collide. The kinetics of primary recrystallization, as well as crystallization, is defined by the formation rate of centers (the nucleation rate) and the linear growth rate of nuclei (the growth rate). According to Kolmogorov, as well as according to Johnson and Meil, the formal recrystallization kinetics is described by an equation similar to that used to describe crystallization:

$$X = 1 - \exp\left(1 - 0.25\, fG^3\, Nt^4\right),$$

where X is the fraction of the recrystallized volume, f is the coefficient that depends on the nucleus shape, G is the growth rate, N is the number of centers, and t is the time. The formula contains a cubed growth rate G. The number of centers N is raised to the first power. This means that the kinetics of recrystallization is largely determined by the growth rate. Avrami proposed a simplified equation for describing the kinetics:

$$X = 1 - \exp\left(1 - Bt^k\right),$$

where B is a coefficient depending on the degree of deformation and the annealing temperature. In most cases, $1 \leq k \leq 2$. The formal description leaves aside information about the physical mechanisms of nucleation and growth of new grains.

For an understanding of the nucleation process during recrystallization, it is useful to compare this process with that during polymorphic transformation upon heating. The differences are as follows. During recrystallization, nuclei arise discontinuously and are an order of magnitude larger in size than during polymorphic transformation. When transformed, the nuclei of a new phase contribute to a maximum similarity of the crystal lattices at the initial-phase boundary (Dankov–Konobeevsky principle). During recrystallization, the orientation of the nuclei, as a rule, differs from the orientation of the matrix. The subcritical size nuclei of a new phase are unstable and may disappear, while recrystallization nuclei of any size are always stable.

When recrystallized, the nuclei form along with the rearrangement of dislocations. As a result, a high-perfect structure region emerges; it is surrounded by high-angle boundaries. The proposed nucleation mechanisms are based on models of transforming low-angle boundaries formed during polygonization or existing in a deformed matrix into high-angle boundaries.

In the case of pre-recrystallization polygonization, the issue is solved simply: some high-angle misorientation subgrains in regard to neighboring subgrains are ready-made centers (nuclei) of recrystallization. In the case of stabilizing polygonization, the centers are formed, as a rule, at the grain boundaries. Several mechanisms for the formation of recrystallization centers have been proposed. The Beck–Sperry model offers migration of a local region of the boundary of neighboring grains, caused by varying degrees of their work hardening. The diagram in Figure 5.19 illustrates the migration of such a region with the formation of a protrusion.

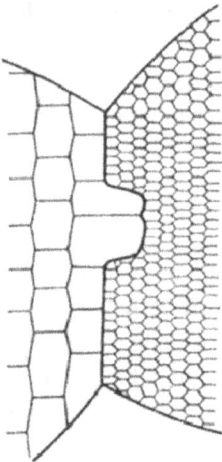

Figure 5.19: Scheme of migration of a section of the boundary due to work-hardening gradient (the Beck–Sperry model).

As a result, a recrystallization nucleus is born. Typically, the region (nucleus) has a size of about 1 μm. With this size, an increase in the grain boundary energy is compensated by a decrease in the bulk energy of work hardening. A model originally advanced by Hu is based on the concept of coalescence of subgrains. The diagram (Figure 5.20) shows the nucleation process of a recrystallization center at the grain boundary by merging subgrains. The merging of subgrains occurs within one of the grains, and growth takes place into another grain that has a high-angle boundary with a nucleus. The Verbraak model applies the idea of the martensitic (shear) mechanism of nucleation when the abrupt emergence of the nucleus is observed.

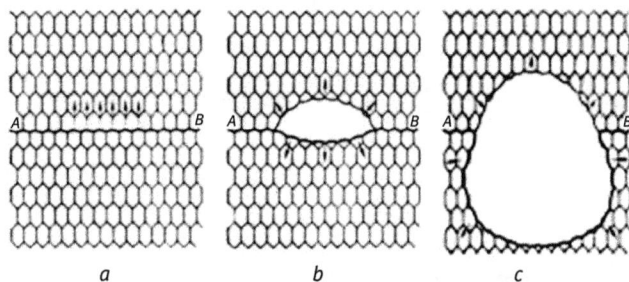

Figure 5.20: Scheme of the formation of a nucleus, including coalescence of subgrains (a) and subsequent migration (b, c). AB is the boundary of grains (the Hu model).

Unlike a melting point, the temperature of the appearance of the first nuclei or the temperature of the onset of recrystallization is not a physical constant. There are many factors that affect this temperature. Among these are the degree of previous deformation, the holding time during annealing, and, to a lesser extent, the heating rate during annealing. The above recrystallization laws reflect these dependencies. Kinetics of aluminum recrystallization at different temperatures shows (Figure 5.21) that the temperature range of recrystallization may decrease by 60 °C with increasing the annealing time (the third law).

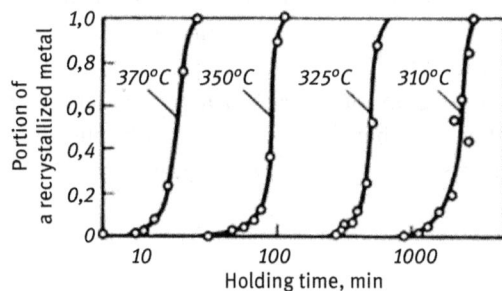

Figure 5.21: Kinetics of recrystallization of 99.96% pure aluminum at different temperatures. Deformation by tension to 10%.

The dependence of the onset recrystallization temperature on the degree of deformation is represented schematically in Figure 5.22.

As the degree of deformation rises, the dislocation density and, accordingly, the bulk energy as the driving force of primary recrystallization (second law) increases.

For an approximate estimate of the onset recrystallization temperature, the Bochvar rule is used, which relates this temperature and the melting point of a given metal:

$$T_{rec} = (0.3 - 0.4)T_{melt}.$$

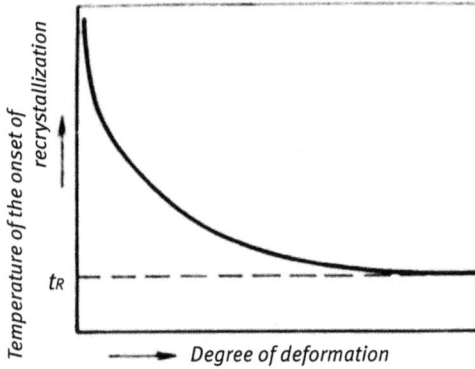

Figure 5.22: Influence of the degree of deformation on the temperature of the onset of recrystallization.

For highly pure metals, the proportionality coefficient drops to 0.25–0.3.

Recrystallized grains grow, owing to the rapid migration of their boundaries into a deformed matrix. When migrating, the high-angle boundary "sweeps out" on its way the lattice defects of the deformed matrix. As a consequence, the volume of the recrystallized structure expands. Boundary migration during recrystallization is the result of the transition of individual atoms or their small groups from a deformed grain to a recrystallized one. In its nature, the mechanism of transition is diffusion. Atoms overcome the potential barrier due to thermal fluctuations.

Misorientation between the new grains and the deformed matrix into which they grow dictates their growth rate. Observations of recrystallization in a deformed aluminum single crystal showed that the nuclei have a maximum growth rate when their misorientation relative to the matrix is approximately 38° (Figure 5.23).

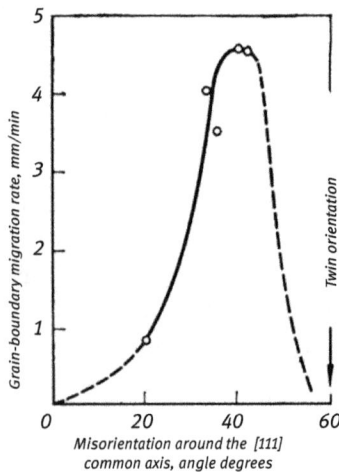

Figure 5.23: Growth rates of nuclei at 615 °C in a single crystal of Al deformed by tension. The [111] axis is common and has one and the same orientation for the nucleus and deformed single crystal.

Such an orientational relationship between the lattices of recrystallized grains and the deformed matrix is common. It gained the name of the Kronberg–Wilson relation. These experiments demonstrated that the low-angle boundaries and the boundaries close to twinning ones have zero or very low mobility (Figure 5.23). Obviously, the mobility of the boundaries is closely related to their structure. At certain angles of misorientation of neighboring grains, atoms at the boundary occupy nodes common for both grains. Such atoms constitute a lattice of coincident nodes. If 1/3 of the nodes coincide, the lattice of coincidence is designated as Σ3. If 1/17 of the nodes coincide, the lattice is denoted as Σ17 and so on. The greater the number of coincident nodes, the lower the boundary energy is. Consequently, the boundary mobility, diffusion along the boundary, and segregation of impurities diminish. Current knowledge about the structure of the grain boundaries can be found in the special literature. Here, it should be noted that orientational effects such as the Kronberg–Wilson moving boundaries do never emerge in the event of being a significant amount of impurity in a solid solution. This is because the impurity powerfully prevents the migration of boundaries. However, a high degree of purity may preclude the orientational effects.

The result of primary recrystallization is a formed microstructure in which the grain size depends on several factors. Among these are the degree of deformation and the recrystallization temperature.

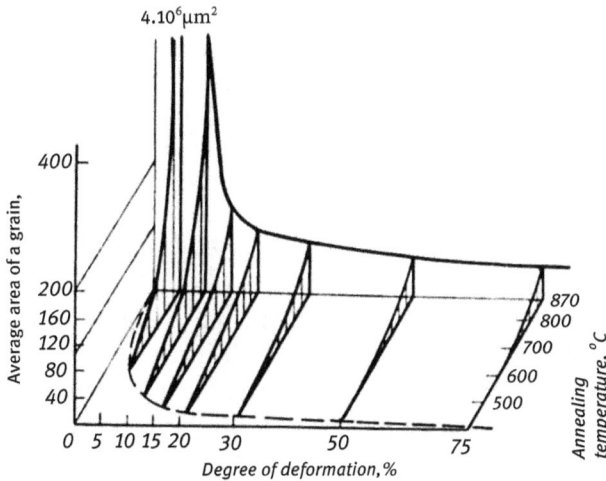

Figure 5.24: Diagrams of recrystallization of electrodeposited Fe. Treatment: preliminary annealing at 930 °C, cold upset, and recrystallization annealing for 1 h.

Three-dimensional recrystallization diagrams (Figure 5.24) depict these dependencies for a fixed annealing time. The vertical cross sections of the diagrams in the axes "average grain area – degree of deformation" show that there is a critical degree of deformation for an appropriate recrystallization temperature, below which the recrystallization does not occur (first law). It can also be seen that the grain size and the recrystallization temperature decrease (fourth law and second law, respectively) as the deformation degree rises. At a critical degree of deformation, the grain size is maximal since the number of recrystallization centers is small, and the growth rate is quite large. The vertical cross sections of the diagrams in the axes "average grain area – annealing temperature" show the grain size is annealing temperature dependent (fourth law). Recrystallization diagrams are used in the technical literature to approximately evaluate the recrystallization temperature range. However, it should be borne in mind that they were built under specific conditions for specific alloys.

5.12 Collective and secondary recrystallization

Once primary recrystallization completed, the grains begin growing with increasing temperature. The grain growth can be uniform (collective recrystallization) and uneven (secondary recrystallization). Figure 5.25 shows a schematic grain distribution in size for the above two cases.

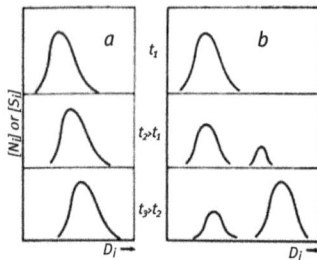

Figure 5.25: Distribution of grains in size during collective (a) and secondary (b) recrystallization: D_i is the grain size, N_i is the number of grains of a given size, S_i is the total area of grains of a given size, t_i is temperature.

With collective recrystallization, the grain distribution shifts toward increasing size as annealing temperature rises, remaining unimodal. Secondary recrystallization enlarges some grains, and the distribution becomes bimodal. During secondary recrystallization, the number of coarse grains and especially their total area increases.

The driving force behind the collective recrystallization is the tendency of the grain boundary energy to reduce. This is because of diminishing the total boundary surface and leveling the surface tension at the grain junctions. A two-dimensional scheme of migration of the boundaries is presented in Figure 5.26.

Coarse grains grow at the expense of small ones. The smaller the grain size, the faster their mergence is. The grain boundaries move to the centers of their curvature, and the angles at the junctions of three grains approach 120°. In this case, an equilibrium configuration is achieved. For simple lattices, the number of edges in the grain section tends to be six. In the three-dimensional case, the equilibrium

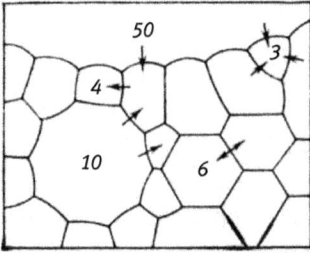

Figure 5.26: Two-dimensional scheme of migration of the boundaries during collective recrystallization. The arrows indicate the direction of migration, and the digits are the number of neighboring grains.

conditions at the grain junctions are somewhat more complicated. In the plane of the polished section, the sizes of arbitrary grain cross sections are different.

The driving force of grain growth during secondary recrystallization can be either bulk energy or grain boundary energy or surface energy, depending on the structural state produced by primary recrystallization. Secondary recrystallization almost gives rise to no grain growth throughout a material but only to the growth of individual grains. In this case, the matrix is stabilized. The size of full-grown grains may exceed the size of the matrix grains by an order of magnitude (Figure 5.27).

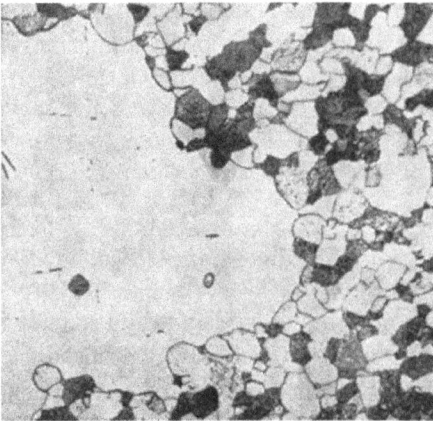

Figure 5.27: A coarse grain fully grown during secondary recrystallization in Zn; ×60 magnification; annealing at 200 °C.

A prerequisite for secondary recrystallization is to stabilize the matrix. One of the reasons for stabilization may be dispersed particles and/or segregation of impurities at the boundaries. This phenomenon is named impurity inhibition of the boundaries. The other reason may be the texture of the primary recrystallization when the grain boundaries are low angle, and thus they are slow-moving ("texture inhibition"). The reason for the preferential growth of individual grains is probably their large size and their higher degree of perfection. A special case of secondary recrystallization is the growth of the centers that form a weak textural component, whereas the stabilized matrix has a different texture of primary recrystallization. It is obvious that, in this case, a crystallographic texture should arise during the secondary recrystallization.

5.13 Recrystallization textures

The texture of primary recrystallization is genetically related to deformation texture and can either inherit it or be different but have an orientational relationship with it, for example, according to the Kronberg–Wilson relation. However, recrystallization texture often does not occur, although the deformation texture did take place. These kinds of texture can be formed depending on the chemical composition of alloys, the presence of impurities, and on the conditions of deformation and annealing, but mainly on the degree of deformation. An example of the sharp texture of the primary recrystallization is cubic texture in thin sheets of magnetic soft alloys such as permalloy (Fe–78% Ni and Fe–45% Ni), having an FCC structure. In the plane of the sheet and in the rolling direction, the (100) plane and the [001] direction are established, respectively. Such texture provides the best magnetic properties of permalloy (squareness of the hysteresis loop).

The texture of secondary recrystallization is genetically related to primary recrystallization texture. A practically important example of the formation of secondary recrystallization texture is cube-on-edge texture or the Goss texture in transformer steel sheets (Fe–3% Si), which has a BCC structure. In the plane of the sheet and in the rolling direction, the (110) plane and the [001] direction are established, respectively. The sharp texture of Goss provides minimal magnetic hysteresis losses during a magnetic reversal, which is an important characteristic in the operation of transformers.

5.14 Dynamic recrystallization

Recrystallization discussed above is called static as opposed to a dynamic one, which takes place during plastic deformation. In this case, plastic deformation is defined as hot. During such deformation, the processes of hardening and softening occur in parallel. The former is caused by an increase in the density of dislocations, and the latter is due to their redistribution, annihilation, and a decrease in the density. Dynamic recrystallization underlies the hot working of metals by pressure (rolling, forging, stamping, etc.), since softening during hot plastic deformation makes it possible to obtain large deformations without destruction. A detailed study of dynamic recrystallization at different temperature and time parameters is the scientific basis for a new technological process such as controlled rolling.

Nuclei during dynamic recrystallization are formed in the same way as during a static one, but the growth process is different. Initially, the nuclei grow rapidly, but the continuing deformation constantly increases the density of dislocations inside the growing grains. Therefore, the driving force of migration of the boundaries gradually weakens. Having reached their limiting size, the grains cease to grow. Then a new cycle of nucleation of grains begins. Figure 5.28 represents schematically a change in the strain rate during creep.

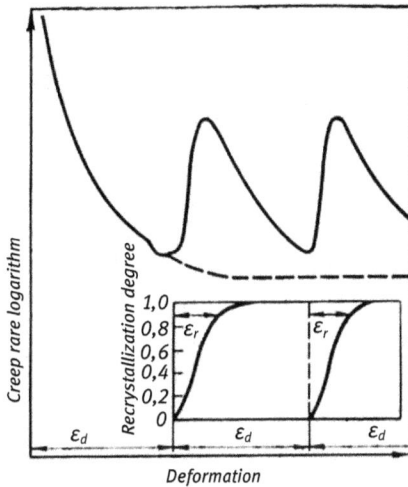

Figure 5.28: Dynamic recrystallization during creep. The relationship between the repetitive recrystallization and the instantaneous creep rate.

Periodically repeating cycles of dynamic recrystallization (see inset in Figure 5.28) are accompanied by an abrupt increase in the creep rate. If the strain ε_d required for the onset of recrystallization is greater than the strain ε_r during recrystallization, that is, $\varepsilon_d > \varepsilon_r$, and if the recrystallization proceeds quickly enough, hardening and softening are separated in time, and the stress–strain curve has a sawtooth shape. If the ratio of deformations is inverse, $\varepsilon_d < \varepsilon_r$, the processes of hardening and softening occur simultaneously in different parts of the deformable metal, and the stress–strain curve is smoothed. Detailed studies of the structure and mechanical properties of steels during hot deformation were performed by Bernstein et al.

5.15 The influence of relaxation and recrystallization on the properties of metals

The mechanical and physical properties of a deformed metal during annealing change in the direction opposite to their change during plastic deformation. Strength characteristics (yield strength, temporary resistance, and hardness) decrease, approaching those in a nondeformed state. Ductility characteristics (elongation, lateral compression) increase. The change in mechanical properties is mainly due to a change in the dislocation structure. Therefore, strength lowers and plasticity rises chiefly during primary recrystallization. With pre-recrystallization annealing (with relaxation), the change in these properties is substantially less.

Latent work-hardening energy, as mentioned in Section 5.7, is released in several stages, each of which corresponds to a specific type of defect. At the first stage, point defects such as interstitial atoms and vacancies are annealed out. The second stage initiates a transformation of the dislocation structure. The third stage includes the recrystallization. Separation of the first two stages is difficult to bring about and is achieved only in the case of deformation at cryogenic temperatures. This is because the point defects can be annealed at temperatures below room temperature. Most of the latent work-hardening energy is liberated during recrystallization when new grains begin nucleating and growing.

An example of the change in properties upon recovery and recrystallization in deformed industrial pure copper is given in Figure 5.29.

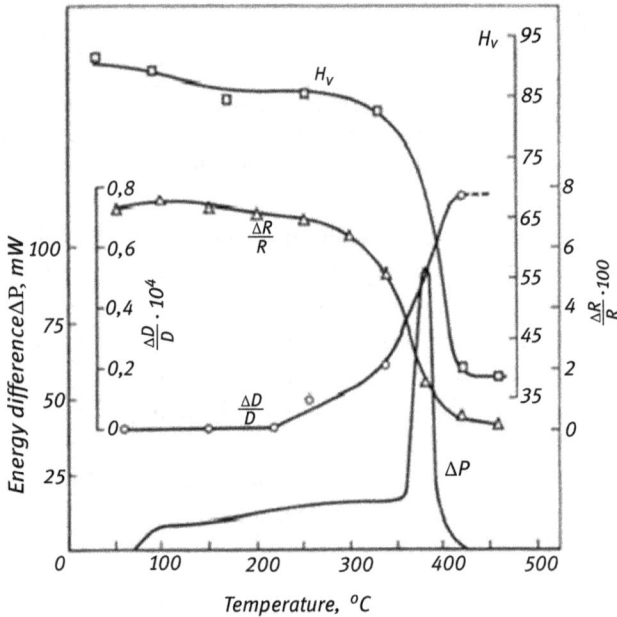

Figure 5.29: Release of stored energy ΔP, changes in hardness H_v, electrical resistance R, and density D during annealing of deformed copper. Deformation by 33% tension at room temperature.

Energy release begins at about 70 °C (with relaxation) and occurs gradually (without separation in the stages) until a sharp peak appears at ~ 370 °C, which corresponds to recrystallization. The recrystallization also contributes to the main decrease in hardness, although a slight decrease is observed at low temperatures. Diminishing in electrical resistance repeats the behavior of hardness. It should be noted that the effect of reducing electrical resistance, associated with the elimination of defects, is negligible in the case of taking electrical resistance measurements at room temperature. In that event, electron scattering by phonons neutralizes this effect. Measurements at cryogenic temperatures (helium temperatures) minimize scattering by phonons and make it possible to more clearly ascertain this effect. The density slightly increases during recovery and recrystallization. The effect is mainly associated with the annealing of vacancies and vacancy clusters.

If a texture arises during recrystallization, the anisotropy of some properties is observed. Examples of anisotropy of the magnetic properties due to texture are given in Section 5.13. Inherent in the deformed state, the anisotropy of the mechanical properties may persist after recrystallization. Such anisotropy is associated not only with the crystallographic texture but also with the texture in the arrangement of foreign particles such as nonmetallic inclusions in steel, oxides, carbides, and so on.

The above section discussed patterns of changes in the properties of deformed metals during relaxation and recrystallization without phase transformations during annealing. If the annealing leads to the decomposition of a supersaturated solid solution or other transformations, the total result of the change in properties may change greatly.

Chapter 6
Phase transformations in the solid state

6.1 First- and second-order phase transitions

The ultimate goal of phase transformations is to obtain materials with desired physical and mechanical properties. Research of the transformations of different types and emerging structures in the chain "heat treatment–phase transformations–structural changes–properties" play a decisive role in elucidating their general regularities or analyzing them in specific alloys. Both ways are implemented in scientific practice.

A thermodynamic approach to the analysis of phase transformations in the solid state is similar to that of crystallization. However, there are two important features determined by the solid state. First, a change in the crystal lattice during the transformation leads to the appearance of elastic strain energy not observable during crystallization. As a rule, the volume of the substance also changes, which also gives rise to elastic stresses. A change in free energy is written as follows:

$$\Delta\Phi = (F_2 - F_1)\,V + \sum \sigma_i S_i + (\partial E_{\text{elas}}/\partial V)\,V, \qquad (6.1)$$

where F_2 and F_1 are the chemical free energies of the final and initial phases in the volume V, $\sum \sigma_i S_i$ is the energy of the interphase surfaces formed, and $(\partial E_{\text{elas}}/\partial V)\,V$ is the energy of elastic deformation. Second, the diffusion mobility of atoms in the solid state is significantly less than in the liquid one. As a result, there arises the possibility of cooperative motion of large groups of atoms. All of the above-said refers to phase transitions of the first order.

Solid state physics considers phase transitions of first and second orders. For a first-order phase transition, a jump of the first derivatives of the thermodynamic potential is typical (e.g., specific volume). A second-order phase transition is accompanied by a jump in the second derivatives of the thermodynamic potential (e.g., the temperature expansion coefficient), whereas the first derivatives change smoothly.

The first-order phase transition generates a number of extremely small regions of the new phase with properties different from those of the initial phase, that is, a new phase nucleates. Nucleus arises, as a rule, by defects in the crystal structure, such as grain boundaries and dislocations. As in crystallization, this nucleation way is called heterogeneous. However, homogeneous nucleation in defect-free places of the crystal lattice is not excluded. The mechanisms of growth of nuclei in different transformations may be different. For example, at martensitic transformation, the coordinated movement of large groups of atoms causes the growth of a new phase, whereas at decomposition of supersaturated solid solutions, the attachment of individual atoms occurs. The second-order phase transition produces small changes in large volumes

https://doi.org/10.1515/9783110758023-006

of the initial phase. The transition is brought about by the gradual accumulation of changes in the entire volume.

The dependence of free energy on a certain thermodynamic parameter during the second-order phase transition is shown in Figure 6.1a.

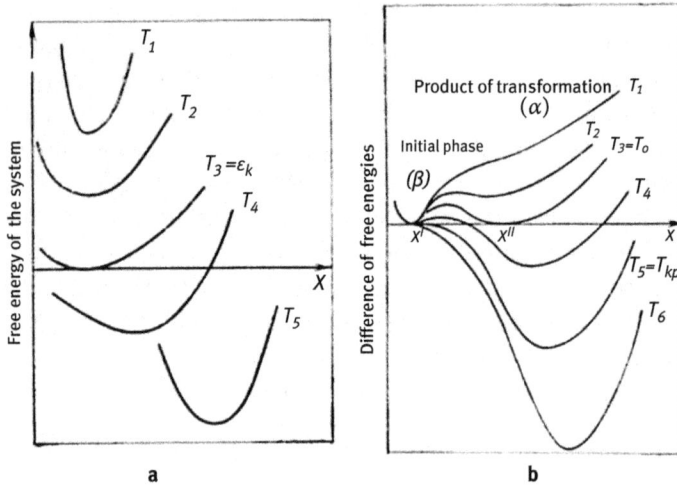

Figure 6.1: Dependencies of free energy on a thermodynamic parameter for second-order (a) and first-order (b) transitions.

At temperatures T_1 and T_2, the initial high-temperature phase is stable. The transformation begins at a temperature T_3 (at the Curie point θ_K) and continues at T_4. At temperature T_5, a new low-temperature phase is stable. The change in free energy during the first-order phase transition is depicted in Figure 6.1b. At temperatures T_1 and T_2, the high-temperature β-phase is stable. At $T_3 = T_0$, the free energies of the initial β-phase and the resulting α-phase are equal. However, the transformation does not begin, since the energy barrier associated with the formation of the interfacial surface and elastic resistance of the medium must be overcome. At a temperature of $T_5 = T_k$, the barrier becomes equal to zero. In the interval $T_0 > T > T_k$, the α-phase or metastable phases toward a stable α-phase are possible to nucleate in a fluctuation way. At temperatures of $T \leq T_k$, barrier-free growth of the α-phase is probable.

This chapter deals with some first-order transitions, namely, martensitic transformations, decomposition of supersaturated solid solutions, and ordering. Various transformations in steels will be considered in Chapter 7.

6.2 Martensitic transformations

The first definition of martensite is associated with the study of quenching of steel. Martensite is a supersaturated solid solution of carbon in α-iron, which has a body-centered cubic (BCC) structure, obtained by rapid cooling from the face-centered cubic (FCC) γ-region (Figure 7.1). The solubility of carbon in γ-iron is greater than in α-iron. Once rapid cooled, the FCC crystal lattice undergoes rearrangement into the BCC structure; carbon is not released, creating a significant supersaturation. In the process, the BCC lattice transforms into a body-centered tetragonal (BCT) lattice whose tetragonality depends on the amount of carbon. Afterward, martensitic transformations were discovered in many metals and alloys, and the product of martensitic transformation was called martensite. The high-temperature phase in alloys of other metals is often called austenite.

Classical studies of martensitic transformations were carried out by G.V. Kurdjumov and his colleagues. According to him, martensitic transformation consists in a regularly coordinated rearrangement of the lattice; atoms do not change places but are only shifted relative to each other at distances not exceeding the interatomic ones. Rearranging, atoms migrate in certain directions with respect to their neighbors. As a result, a shift occurs. Experimentally, a shift makes itself felt in emerging a characteristic relief (protrusions and depressions) on a flat polished surface (thin section). A quantitative study of the relief yields insight into the nature of the martensitic transformation (A.B. Greninger and A.R. Troiano) and enables one to build up a phenomenological (crystallographic) theory of martensitic transformations (see below).

The main peculiarities of the martensitic transformation are the diffusion-free nature, the presence of a strict orientational relationship between the martensitic and austenitic lattices, the spreading of the transformation over a wide temperature range, the rapid deceleration of the transformation when cooling stops, and the impossibility of delaying the transformation below its onset temperature (martensitic point M_s). Later, an isothermal martensitic transformation was found at low temperatures, the reversibility of martensitic transformations during heating and cooling was established, and the phenomenon of thermoelastic equilibrium during martensitic transformations was discovered. Let us dwell on these features in more detail.

6.2.1 Thermodynamics of martensitic transformations

At martensitic transformation, there is no redistribution of alloy components between the phases; martensite and austenite have the same chemical composition. Therefore, the martensitic transformation can be regarded as a transformation in a single-component system. The martensite and austenite are in thermodynamic equilibrium at a temperature T_0, at which the free energies of the phases are equal (Figure 6.2).

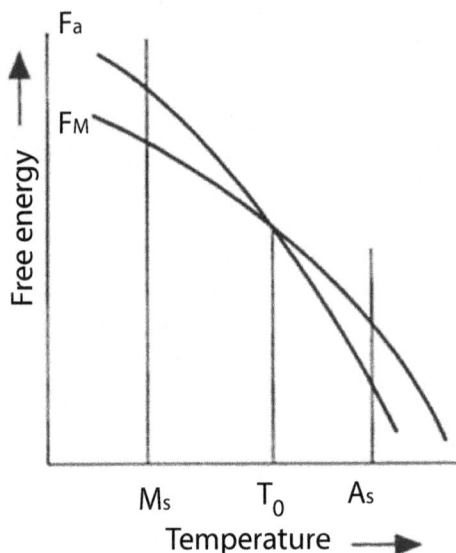

Figure 6.2: Temperature dependencies of free energies for austenite (F_A) and martensite (F_M).

The start of the martensitic transformation (at a temperature M_s) requires super-cooling below T_0 to compensate for an increase in the surface and elastic energies (formula (6.1)). The role of elastic energy in the martensitic transformation is extremely large. In many alloys, martensite is also transformed back into austenite. Some overheating relative to T_0 also provokes the transformation. (An exception is thermoelastic martensitic transformations, see below.) The reverse martensitic transformation to be implemented excludes the diffusion redistribution of components in martensite. For this, the reverse transformation temperatures must be low, and the heating rate must be high enough. The reverse transformation is the reverse one not only because that it occurs upon heating, whereas the direct martensitic transformation occurs upon cooling. The paths of migrating atoms in the reverse transformation can be opposite to those at the direct transformation. In this case, the relief that appears on a flat polished surface is the opposite to the relief at direct transformation.

Given that the temperature T_0 depends on the chemical composition of alloys, G.V. Kurdjumov proposed a simple schematic diagram of direct and reverse martensitic transformations of solid solutions of different concentrations (Figure 6.3).

The scheme illustrates the need for overheating, a wide temperature range of transformations, and a hysteresis of direct and reverse transformations. Figure 6.4 gives an example of the implementation of direct and reverse transformations in a Fe–Ni system.

The temperature range of martensitic transformations in iron–nickel alloys is stretched by ~ 100 °C; the hysteresis of the direct and reverse transformations is

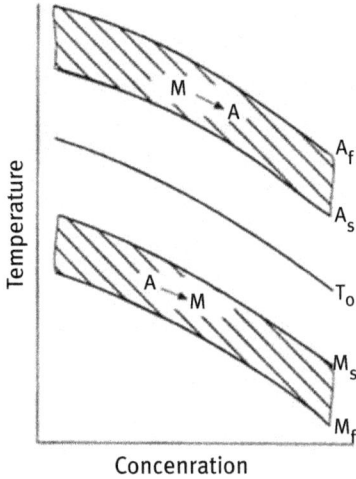

Figure 6.3: A schematic diagram of direct and reverse martensitic transformations of solid solutions for different concentrations.

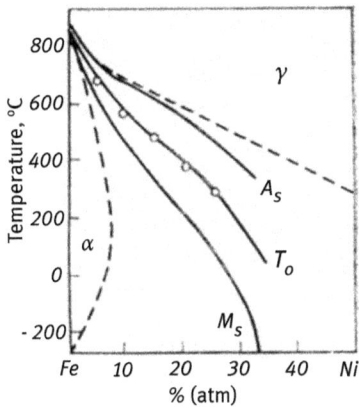

Figure 6.4: Temperature dependencies of the start of martensitic transformation (M_s) upon cooling, reverse martensite transformation (A_s) upon heating, and the equality of free energies of austenite and martensite (T_0) on the content of Ni in a Fe–Ni system. Dashed lines are the boundaries of α- and γ-regions on the equilibrium diagram.

several hundred degrees. The temperature T_0 at which the free energies of martensite and austenite are equal is calculated by the formula:

$$T_0 = 1/2\,(M_s + A_s).$$

The position of temperatures T_0 and M_s in a Fe–C system is shown in Figure 7.10. The temperature difference $T_0 - M_s$ reaches 200–250 K.

6.2.2 Kinetics of martensitic transformations

The kinetics of phase transformation is determined by the rate of nucleation of a new phase and the rate of their growth. At martensitic transformation, the conjugation of the crystal lattices of martensite and austenite is coherent or partially coherent. This circumstance provides high mobility of the interphase boundary. Therefore, the growth rate of martensite crystals is extremely high and reaches a magnitude of the order ~ 1 km/s in iron alloys. Hence, the kinetics of the martensitic transformation is limited by the nucleation rate. The nucleation rate (the number of nuclei arising in a unit volume per unit time) is written as the product of two exponentials:

$$N = A \exp\left(-W/kT\right) \exp\left(-U/kT\right),$$

where W is the work of formation of a nucleus of critical size and U is the activation energy of the transition of atoms from austenite to the nucleus. Under conditions of coherent growth, the activation energy U is small, and the nucleation rate is controlled by the first exponent. The work of formation of a critical size nucleus W diminishes with increasing supercooling ΔT according to the law

$$W = K/(\Delta T)^2,$$

where K is a constant that depends on many factors. For large values of ΔT, the value of W tends to zero, and the transformation becomes barrier free.

According to the theory of homogeneous nucleation, we have

$$\log N = 28 - K/(\Delta T)^2.$$

Using this expression, it is easy to calculate that an increase in supercooling by 10% heightens the density of nuclei by 10^6 times. This means that rapid cooling makes the nucleation rate and the growth rate of the nuclei be very high during their coherent growth. In this case, the martensitic transformation occurs so quickly that it cannot be experimentally detected during isothermal holding, that is to say, it is athermal.

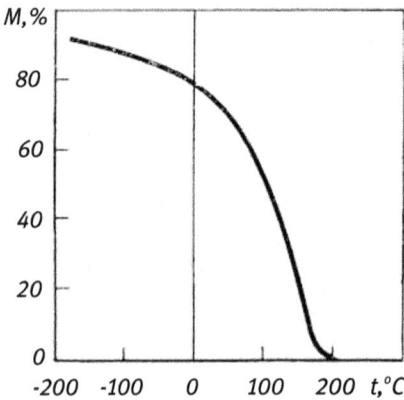

Figure 6.5: A curve of athermal martensitic transformation.

Figure 6.5 displays an increase in the amount of martensite during continuous cooling in the temperature range of M_s–M_f (martensitic curve) for athermal martensitic transformation. When cooling stops, the transformation ceases but resumes during further cooling. A special case of athermal transformation kinetics is explosive martensitic transformation, which is typical of iron alloys with a low martensitic point (see Figure 7.11). Explosive kinetics manifests itself in another feature of the martensitic transformation – autocatalyticity. In other words, the emergence of a single martensite crystal causes a chain reaction to form a large group of crystals, with the transformation being accompanied by sound effects. Explosive kinetics is due to a high level of internal stresses arising at the edges of martensite crystals.

However, since the martensitic transformation is a "usual" first-order phase transition, which is implemented by the nucleation and growth of a new phase, this transformation should be expected to proceed according to the "normal" isothermal kinetics with an incubation period. G.V. Kurdjumov and O.P. Maksimova discovered such a transformation in highly alloyed iron alloys. Figure 6.6 sketches a change in the amount of martensite during isothermal holdings at different temperatures in a Fe–Ni–Mn alloy.

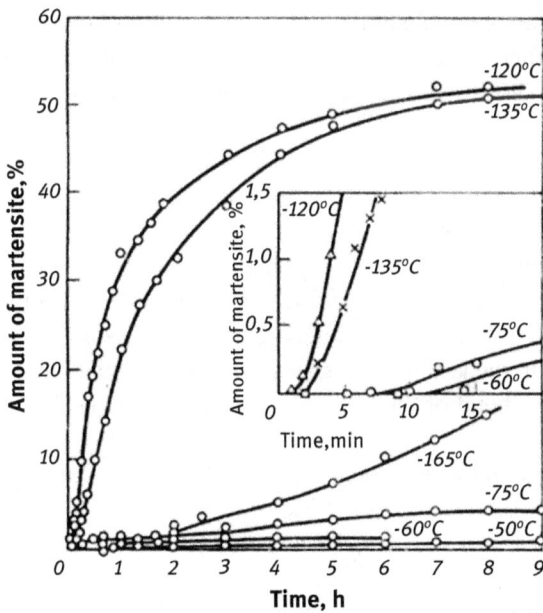

Figure 6.6: Isothermal martensitic transformation at different temperature for a Fe–23.4%Ni–3.3%Mn alloy.

At temperatures of −120 and −135 °C, the transformation occurs more completely than at −165, −75, and −60 °C. Figure 6.7 presents C-shaped curves of the initial stages of isothermal martensitic transformation in this alloy. Once discovered, the completely isothermal transformation showed that the nucleation of martensite

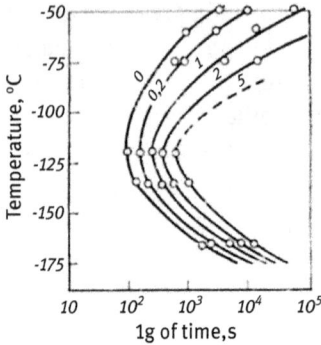

Figure 6.7: C-shaped curves of martensitic transformation for a Fe–23.4%Ni–3.3%Mn alloy. The curves' numbers indicate the amount of martensite.

crystals is a temperature-dependent, thermally activated process as much as any process of nucleation of crystals of a new phase. However, nucleation work W and activation energy U are substantially less during the martensitic transformation compared with those at crystallization or diffusion phase transformation. This stands for a high rate of the martensitic transformation. It should be underscored that a certain amount of martensite is often isothermally formed after athermal or explosive transformation. Also, athermal transformation can occur after isothermal one. Therefore, there is no sharp difference between the athermal and isothermal kinetics.

G.V. Kurdjumov came to the conclusion that there is a coherent bond between the crystal lattices of martensite and austenite. This follows from the Kurdjumov–Sachs (KS) orientation relationships (ORs), as well as from the fact that the growth rate of martensitic crystals is extremely high. There are two possible cases of the growth of martensitic crystals: with loss of coherent bond and without loss. If the coherent bond is broken after the crystal reaches a certain size, the latter's growth almost stops. This is because the activation energy U in the lack of coherence increases by more than an order of magnitude. If the coherent bond of the lattices is preserved, thermoelastic equilibrium is possible to take place.

We write down the total change in free energy ΔF upon the formation of a martensite crystal, similarly to (6.1):

$$\Delta F = -\Delta F_{\text{bulk}} + E_{\text{surf}} + E_{\text{elas}},$$

where ΔF_{bulk} is the "driving force" of the transformation, proportional to the volume of the crystal obtained; E_{surf} is the surface energy, which can be neglected during coherent growth; E_{elas} is the elastic energy. During the growth of the martensitic crystal upon cooling, E_{elas} increases faster than ΔF_{bulk}. When the condition $E_{\text{elas}} = \Delta F_{\text{bulk}}$ holds, the driving force is $\Delta F = 0$, and the crystal stops growing. During further cooling ($\Delta F_{\text{bulk}} > E_{\text{elas}}$), the crystal growth resumes until the equilibrium $E_{\text{elas}} = \Delta F_{\text{bulk}}$ is established again. If the temperature rises (heating, $E_{\text{elas}} > \Delta F_{\text{bulk}}$), the crystal declines in size and the elastic equilibrium of the martensitic crystal and matrix (austenite) changes. Therefore, it can be said that **thermoelastic equilibrium** is reached. The phenomenon of thermoelasticity was predicted by G.V. Kurdjumov and experimentally confirmed by Kurdjumov and Khandros. Figure 6.8 depicts this phenomenon: the growth of a martensitic crystal upon cooling and its contraction upon heating in a Cu–Al–Ni alloy.

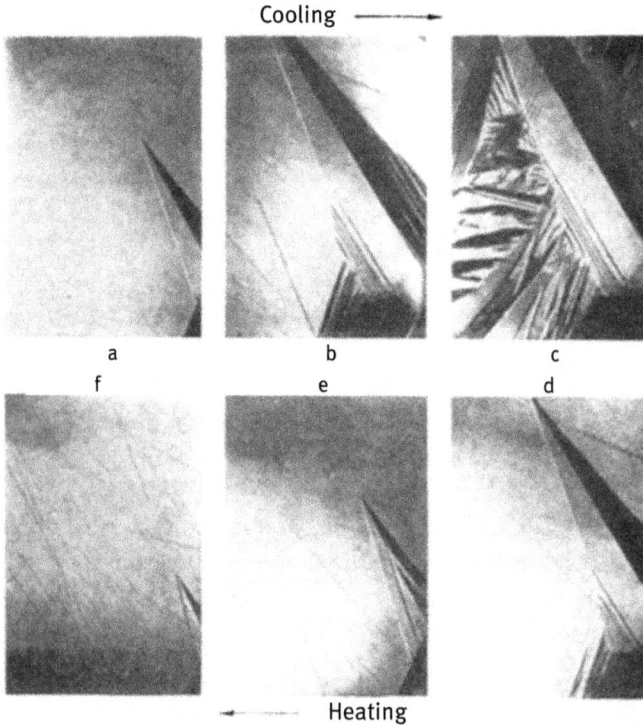

Figure 6.8: Growth of a thermoelastic martensite crystal during cooling (a–c) and its reduction during heating (d–f) in a Cu–Al–Ni alloy. The magnification is 60.

It should be noted that the thermoelastic behavior produces the reverse martensitic transformation at temperatures below T_0 and almost without a hysteresis (or with a slight hysteresis of several degrees). The martensite/austenite interphase boundary shears in the opposite direction and the atoms appear to migrate along the same paths as at the direct transformation but in the opposite direction. The phenomenon of thermoelasticity underlies the effects of shape memory and superelasticity (see below). The implementation of the thermoelastic martensitic transformation is guaranteed by the accumulation of high elastic energy and the absence of relaxation processes that violate coherence at the interface.

6.2.3 The problem of nucleation at martensitic transformations

The issue of the nature of martensite nuclei is far from being resolved. Homogeneous nucleation is usually excluded, since the work of homogeneous formation of a critical nucleus is huge due to large values of elastic energy. The heterogeneous nucleation of martensite at the boundaries and sub-boundaries has not been experimentally detected. The nucleation centers may be dislocation structures that approximate the configuration of a given region of the matrix to the structure of the nucleus. Another interpretation argues that dislocations create stress fields that reduce the work of the formation of a nucleus up to barrier-free nucleation.

The problem of nucleation deals with the issue of pre-martensitic states (V.G. Pushin). Some alloys (in particular, titanium nickelide) exhibit subtle structural changes at temperatures near the temperatures of the onset of martensitic transformation. These changes precede the formation of martensite and are interpreted as a loss of the mechanical stability of the austenite lattice. It remains unclear whether such behavior is random for individual alloys or it bears a general nature.

6.2.4 Phenomenological crystallographic theory of martensitic transformations

As said earlier, there is an orientational connection between the lattices of martensite and austenite. ORs are usually written through the parallelism of the planes and directions lying in these planes. Let us look into the transformation of austenite to martensite in iron alloys. For this, we select a tetragonal cell in an FCC lattice of austenite, as shown in Figure 6.9.

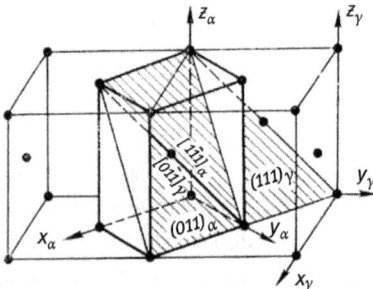

Figure 6.9: Preimage of a BCC cell (α) in an FCC lattice (γ).

According to the Bain scheme, this cell is a prototype of the unit cell of a BCC (or BCT) martensite lattice. From Figure 6.9, it follows that the prototype of the martensite lattice and the austenite lattice is related to the OR in the following way:

$$(011)_\alpha \big\| (111)_\gamma \quad \text{and} \quad [1\bar{1}1]_\alpha \big\| [0\bar{1}1]_\gamma.$$

These ORs are known as KS relationships. However, in order to obtain a BCC lattice, it is necessary to perform the Bain distortion: to compress a tetragonal cell by 20% in the [001] direction and stretch it by 12% in the [100] and [010] directions (Figure 6.10).

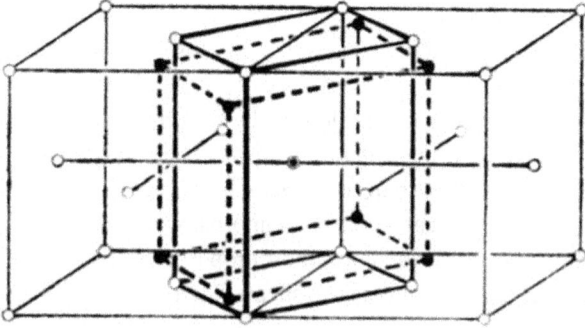

Figure 6.10: The Bain deformation. Preimage of a BCC cell is shown with solid lines. Dashed lines indicate a BCC cell.

In this case, the KS ORs are violated. For their recovering, the BCC lattice should be rotated by 10° (Figure 6.11).

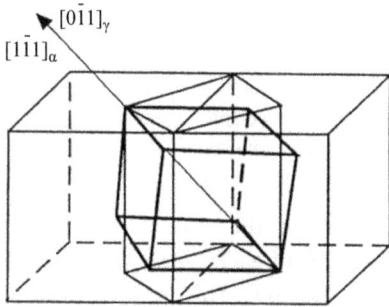

Figure 6.11: A BCC-cell rotation needed for the implementation of the K-3 orientation relationship.

The experimental determination of the ORs has been the subject of numerous studies. For steels with high carbon content (0.5–1.4%), the KS ORs were found. These ORs are the best alignment to the Dankov–Konobeyevsky principle (see Chapter 4): close-packed planes and close-packed directions are mutually parallel. An FCC austenite lattice contains four crystallographically equivalent planes of the {111} type and six crystallographically equivalent directions of the <112> type. Therefore, for 1 austenite crystal, 24 orientations of martensite crystals are possible, which satisfy the KS orientational relationships.

For iron–nickel alloys with high nickel content (28–32%), Nishiyama derived the following ORs:

$$(011)_\alpha \| (111)_\gamma \quad \text{and} \quad [0\bar{1}1]_\alpha \| [2\bar{1}\bar{1}]_\gamma.$$

They differ from the KS ORs by a rotation of 5.26° around the normal to the $(111)_\alpha$ plane. The Nishiyama ORs read that 12 orientations of martensite crystals are possible for 1 austenite crystal. An accurate determination of these ORs indicates that for medium-carbon steels, ORs that intermediate between the ORs of KS and Nishiyama are usually realized.

The Bain deformation gives a visual representation of the transformation of the lattice during the martensitic FCC→BCC transformation. However, the large magnitude of compression and tensile strains and the need to rotate by a large (~10°) angle make such a restructuring path unrealistic. Let us consider another way of transforming the lattice at the martensitic FCC→BCC transformation proposed by KS. The restructuring of the lattices is described by two successive shears (Figure 6.12).

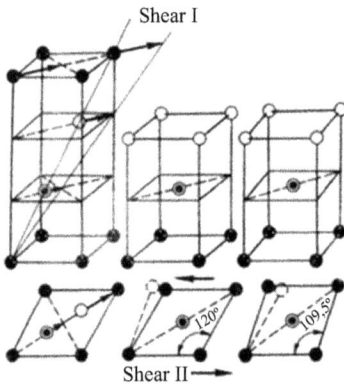

Figure 6.12: A scheme of a two-shifted restructuring of an FCC–BCC lattice by Kurdjumov–Sachs.

The geometry of the first shear is based on the similarity of the martensitic transformation with twinning and consists in a twin-like shear of each plane $[0\bar{1}1]_A$ in the $(-211)_A$ direction by 1/6 of the large diagonal of the rhombus. This shear swaps the sequence of laying close-packed *ABCA* planes for *ABA*. To obtain a BCC lattice, a second small shear and some change in size are required. In the $[0\bar{1}1]_A$ direction, the second shear in the $(-211)_A$ plane perpendicular to $(111)_A$ leads to a change in the angle between the close-packed directions from 120° to 109.5°. The combination of the described strains creates a BCC lattice related to the initial FCC lattice of the KS ORs. The given way of restructuring is more realistic than the Bain deformation with rotation.

Further development of the concepts of martensitic transformation is associated with the determination of the shape deformation: a habit plane, direction, and magnitude of the macroscopic shear. A.B. Greninger and A.R. Troiano conducted a quantitative research of the relief of a Fe–22%Ni–0.8%C alloy, which arises on the polished surface during martensitic transformation. A schematic representation of the relief is given in Figure 6.13.

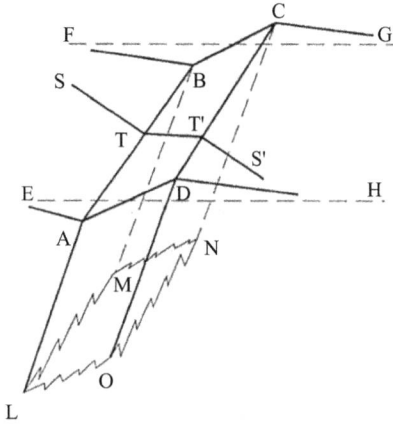

Figure 6.13: Deformation of the shape and the appearance of the relief in the plane surface to obtain a martensite plate.

As a result of the macroscopic shear, the *ABCD* surface of the martensitic crystal inclined relative to the EFGH initial surface, remaining flat. Straight lines (e.g., SS') experience kinks, saving continuity. The *TT* segment keeps a straight line. The interphase surfaces *ABML* and *DCNO* do not rotate or distort; they are only slightly shifted in their planes. Therefore, macroscopic shear deformation (i.e., shape deformation) is a deformation with an invariant (habit) plane. Measurements of the magnitude of the shear showed that the shape deformation is significantly less than the lattice deformation that ensures the implementation of the ORs. Now, it became clear that it is necessary to enter an additional deformation to ultimately generate a shape deformation, without violating the ORs. With an invariant lattice, such deformation can be brought about by sliding or twinning.

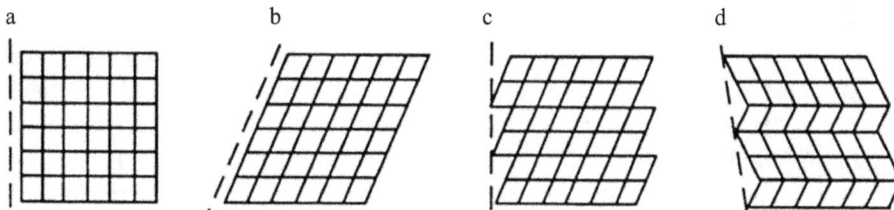

Figure 6.14: The types of deformation at martensite transformation: (a) an initial crystal, (b) deformation of the crystal lattice, (c) invariant-lattice deformation by sliding, and (d) invariant-lattice deformation by twinning.

Figure 6.14 schematically demonstrates strains during martensitic transformation: lattice deformations (b) and shape deformations in which lattice deformation is combined with invariant-lattice deformation by sliding (c) and twinning (d). Electron microscopic observations of twins and organized dislocation structures inside martensitic crystals confirm the concept of the phenomenological theory of martensitic transformations about invariant-lattice deformation.

The phenomenological theory of martensitic transformation, proposed by Wechsler, Liebermann, and Reed presents the shape deformation P_1 as the product of three matrices:

$$P_1 = RBP,$$

where R is the rotation of a solid body, B is the Bain deformation, and P is the deformation of an invariant lattice (simple shear deformation). The Bain deformation matrix has the form

$$B = \begin{pmatrix} \eta & 0 & 0 \\ 0 & \eta & 0 \\ 0 & 0 & -\eta_1 \end{pmatrix}$$

For iron alloys, $\eta = 1.12$ and $\eta_1 = 0.80$.

The invariant-lattice deformation matrix is determined by the method and parameters of this deformation and is specified for each martensitic transformation.

It is worth noting that the lattice deformation and the invariant-lattice deformation are inseparable and represent a united process in reality. General deformation of the shape minimizes elastic stresses in the habit plane (interphase boundary). However, the ends of martensitic crystals preserve a high level of stresses. The latter are due to the autocatalyticity of the martensitic transformation, as well as the choice of the orientation of the neighboring crystal relative to the crystal earlier obtained.

A.L. Roytburd generalized the theory of martensitic transformations in his works and determined methods for minimizing elastic stresses, which play a crucial role in all aspects of martensitic transformations. The tendency of a system to minimize its free energy is manifested in the following: (a) strict adherence of the orientational relationships between lattices; (b) a more or less strict similarity of martensitic crystals in shape to a plate; (c) more or less strict orientation of the plate's habit plane relative to the crystal lattice of the initial phase; (d) domainization of a martensitic crystal to form twinned orientations; (e) plastic stress relaxation, especially at the edges of the martensitic plate; and (f) ordering of the choice of specific orientational relationships with the corresponding planes and directions of macroscopic shear by neighboring martensitic crystals.

6.2.5 Microstructure of martensite

The minimization of elastic stresses determines the formation of a martensitic structure: the shape of crystals, ways of their grouping, and their internal structure. The formation of an individual martensitic thin-plate-shaped crystal is associated with the least elastic (and surface) energy. The combining of the plates into packages, truss-shaped or pyramidal structures, minimizes the elastic energy of a whole group of crystals. Martensite crystals contain twins and stacking faults or have a high dislocation density that confirms the existence of invariant-lattice deformation in the phenomenological theory of martensitic transformation.

The martensite microstructure in steels and iron alloys is set forth in Chapter 7. Martensitic crystals can be divided into four types: lath-shaped (dislocation), butterfly-shaped, lenticular-shaped, and thin-plate-shaped (completely twinned) ones. Each type of crystal usually has a characteristic habit plane.

Martensite crystals in shape-memory alloys contain no dislocations and a high density of twins. Figure 6.15 gives an example of a twinned structure of thermoelastic martensite in a Cu–Zn–Al alloy.

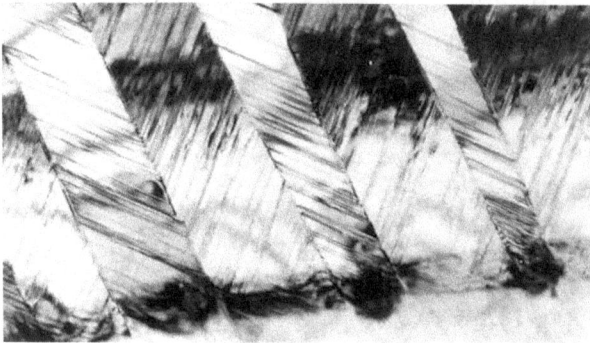

Figure 6.15: Structure of thermoelastic martensite in a Cu–Zn–Al alloy ($\times 30,000$).

The relaxation of elastic stresses to form dislocations violates the reversibility of the martensitic transformation and prevents thermoelastic behavior.

6.2.6 Effects of shape memory. Superelasticity. Reactive stresses

Thermoelastic martensitic transformation underlies the effects of shape memory and superelasticity. As mentioned earlier, thermoelastic transformations (direct and reverse) occur at temperatures below T_0 and have a small temperature hysteresis. Figure 6.16 compares the temperature hysteresis for two alloys: with thermoelastic (AuCd alloy) and non-thermoelastic, athermal transformations (FeNi alloy).

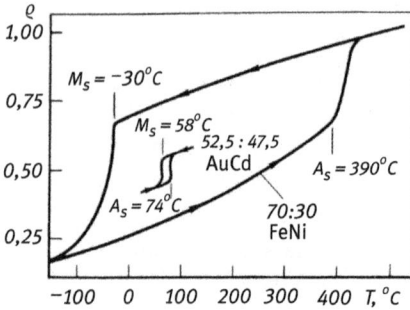

Figure 6.16: Thermal hysteresis at athermal (Fe–Ni) and thermoelastic (Au–Cd) martensite transformations.

The former has a hysteresis of about 15 °C; the latter exhibits it at ~ 400 °C. The reverse transformation of the second alloy occurs above T_0 (Figure 6.4). Figure 6.17 represents schematically overviews of the temperature ranges of the direct and reverse thermoelastic transformations in enlarged scale. The M_s^σ point located above M_s and A_f meets the temperature below which thermoelastic martensite can be obtained under the influence of external stresses. Depending on the deformation temperature, the shape-memory effect or superelasticity can be realized. If stresses are applied in the temperature range $A_f - M_s^\sigma$, superelasticity is observed. A diagram of the superelastic behavior of the alloy with increasing and decreasing external stresses is shown in Figure 6.18.

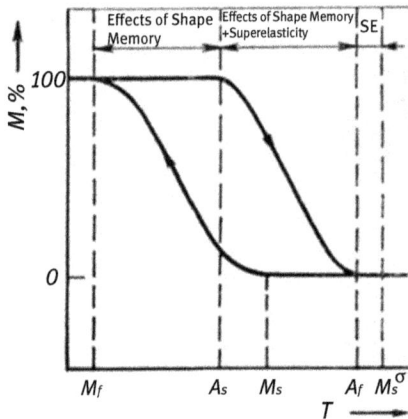

Figure 6.17: Change in the amount of thermoelastic martensite upon cooling and heating.

The segment AB corresponds to elastic deformation of austenite. The segment BC accumulates deformation under constant stress. In practice, a slight increase in stresses exists in this segment. Martensite is formed when exposed to applied stresses. In the process, crystals arise; their deformation is favorably oriented toward the direction of the stresses. Accumulated deformation (deformation of the martensitic transformation) can reach 10%. The segment CD matches elastic deformation of martensite. With

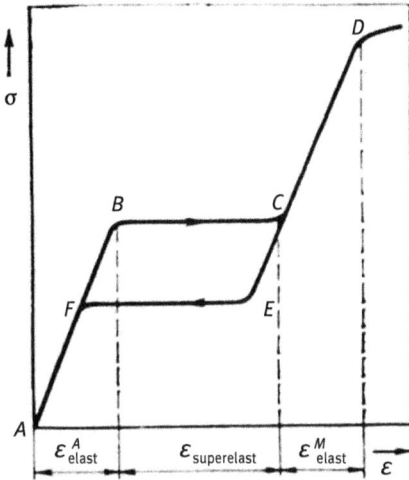

Figure 6.18: Scheme of the superelastic behavior in stress–strain coordinates. ε^A_{elast} and ε^M_{elast} are elastic deformation of austenite and martensite, respectively; $\varepsilon_{superelast}$ is superelastic deformation.

unloading, the accumulated deformation disappears in the *EF* segment due to the reverse transformation because the alloy temperature is above A_f. In contrast to the usual elastic deformation, the superelastic deformation disappears, leaving some stress hysteresis. The usual elastic deformation does not exceed tenths of a percent. However, the superelastic deformation can be more than two orders of magnitude. This fact opens up new possibilities for use in various devices.

If deformation of the martensitic transformation accumulates under the action of external stresses at a temperature below A_s (see Figure 6.17), it does not disappear during unloading. If external stresses are applied at a temperature below M_s, the deformation of the martensitic transformation comes about from not only due to the formation of favorably oriented martensite crystals but also due to the reorientation of those that occurred earlier upon cooling. When heating above A_f, the reverse transformation takes place, and the accumulated deformation fully or partially disappears. That is what accounts for the effect of shape memory. An illustration of the effect is shown in Figure 6.19.

In the *BC* segment, the deformation accumulates; the superelastic deformation is retained upon unloading (*CD* segment). During heating, the deformation recovers in the process of reverse transformation (*EF* segment). Large "superelastic" deformation (up to 10%) and its recovery are time-separated. This property is used in various fields of technology (space technology, aircraft building, and instrument making) and medicine (traumatology, surgery, and orthopedics). Applicable scope of shape-memory alloys is very diverse and continues expanding.

If one obstructs the recovery of deformation during heating (e.g., fixing the sample in a block), "reactive" stresses develop. Their maximum magnitude is limited by the yield strength of the memory alloy. Thus, along with deformation effects (shape-memory effect and superelasticity), shape-memory alloys demonstrate force

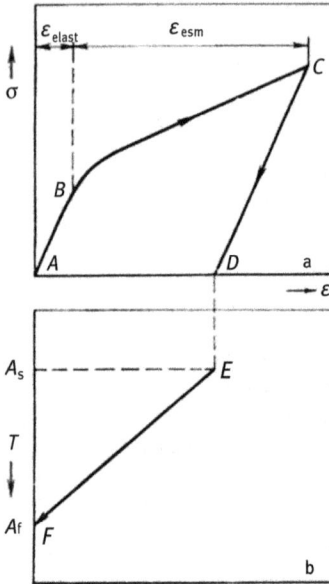

Figure 6.19: A visual representation of the shape-memory effect: (a) a stress–strain diagram and (b) disappearance of deformation upon heating.

effects (stress generation), as well. Figure 6.20 produces a spatial picture of the shape-memory effects and stress generation (V.N. Khachin).

In the stress–strain coordinates, loading diagrams of $\sigma(\varepsilon)$ at two temperatures are shown. At a high temperature, the true yield strength σ_y of austenite is fixed. At a low temperature, the stress σ_M, at which martensite formation begins, is determined. The martensitic shear stress σ_M can be much less than the yield strength σ_y. The deformation–temperature coordinates illustrate the shape-memory effect, $\varepsilon(T)$. The stress–temperature coordinates involve the stress generation and relaxation curves $\sigma(T)$. The shape-memory effect can be repeated multiple times without applying external stresses but under the influence of internal ones. In this case, the shape of the sample (or product) changes reversibly during alternating direct and reverse transformations. This effect bears the name of the reversible shape-memory effect. Figure 6.21 illustrates the manifestation of the reversible shape-memory effect after the deformation of martensite.

The segment 01 corresponds to the initial deformation in the martensitic state. The segment 12 matches the shape-memory effect. The segment 23 demonstrates the reversible shape-memory effect. With the latter, the magnitude of reversible deformation is several times smaller than that upon the shape-memory effect; nevertheless, it is quite large compared to that of conventional elastic deformation. Oriented internal stresses that lead to the selection of martensitic crystal orientations and that create the reversible shape-memory effect can be obtained in various ways: by deformation of martensite or austenite, special training, and aging under stress.

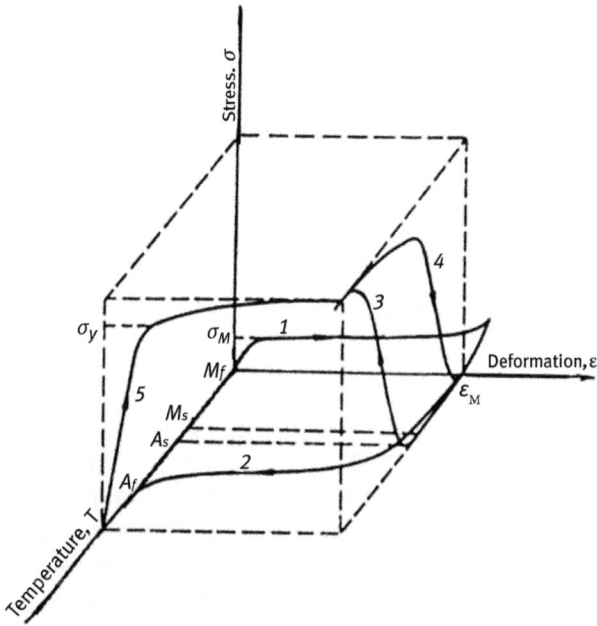

Figure 6.20: A visual representation of martensitic inelasticity (by Khachin): 1,5, stress–strain diagrams; 2, SME; 3, stress generation; 4, stress relaxation.

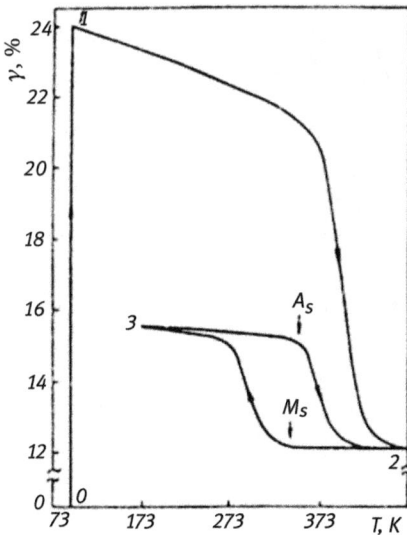

Figure 6.21: A visual representation of the reversible shape-memory effect for a TiNi alloy.

More recently, new Ni_2MnGa-based alloys have been discovered, in which the shape-memory effect is controlled by the action of a magnetic field. Significant deformations (about 7%) exceeding magnetostriction by one or two orders of magnitude can be produced by changing the direction of the magnetic field. In this case, the martensitic crystals are reoriented. A remarkable feature of these alloys consists in interacting the magnetic domain structure with the twin structure of martensite, in other words, with crystallographic domains.

Apart from memory effects, martensitic transformation gives rise to another mechanical effect: an increase in ductility due to the occurrence of the transformation during deformation. Transformation-induced ductility (trip effect) is used in trip steels and will be discussed in Chapter 7.

6.2.7 Massive transformations

In addition to martensitic transformation, there is another type of transformations that occur without changing the chemical composition of phases and deforming the shape of the transformed volume. A new phase comes about from the discordant attachment of individual atoms, that is to say, diffusion over short distances takes place. Diffusion over long distances does not occur; therefore, there is no redistribution of components. For the shape being not deformed, the relief on the polished surface does not appear. The segments of the new phase are single phase; they look like shapeless arrays without any simple border line. Figure 6.22 exemplifies how a massive BCC–FCC transformation proceeds during the heating of an iron–nickel alloy.

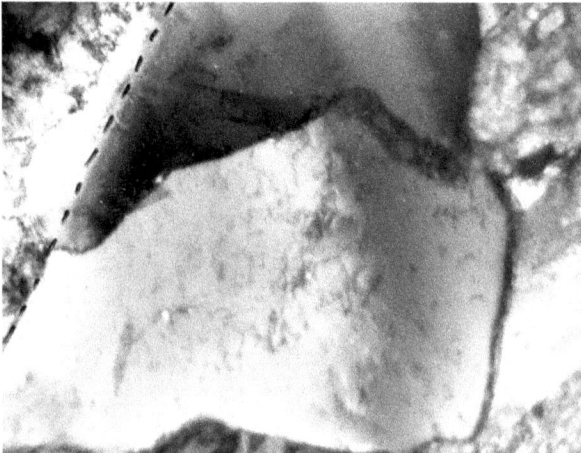

Figure 6.22: Massive BCC–FCC transformation in a Fe–23%Ni alloy($\times 8,000$).

Massive transformations occur either during continuous cooling at a sufficiently high speed that prevents from developing long-distance diffusion or during isothermal holding under sufficiently deep supercooling or during heating. By their nature, massive transformations are intermediate between martensitic and transformations with a change in the chemical composition of the phases.

Massive transformations can be thermodynamically justified by accompanying a decrease in free energy.

Figure 6.23 displays a region of the binary diagram with the mutual solubility of components and a diagram of free energies of two phases. If one holds an alloy of composition c_0 from the single-phase γ-region at a temperature T_1 until equilibrium, decomposition into phases of different compositions occurs according to the reaction $\gamma(c_0) \rightarrow \alpha(c_1) + \gamma(c_2)$. Free energy drops from point a to point d. If one quickly enough transfers this alloy from the γ-region to the same temperature, the reaction $\gamma(c_0) \rightarrow \alpha(c_0)$ may be observed without changing the composition. The free energy lowers from point a to point b. This case stands for a massive transformation.

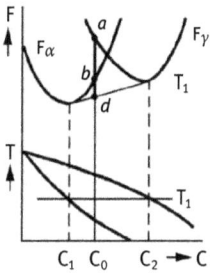

Figure 6.23: A segment of a binary diagram with mutual solubility of components and a scheme of free energies of two phases.

The term "massive transformation" was introduced by Greninger, which is widely used by Massalsky, but is rarely used in Russian literature. Polymorphic transformations in pure metals, if they occur by a non-martensitic mechanism, are massive. At the same time, they are usually called normal. Therefore, in some cases, these terms are equivalent. Examples of massive transformations are transformations in Cu–Al, Cu–Zn, and Fe–Ni alloys upon rapid cooling, when the crystal structure of single-phase solid solutions suffer a change to form shapeless regions. In the process, the chemical composition remains unfailing.

6.3 Decomposition of supersaturated solid solutions

The decomposition of a supersaturated solid solution is phase transformation in which a single-phase solid solution turns into a system of two (or more) phases. The most common case of decomposition processes is the transformation that contributes to the precipitation of an excess phase from the matrix solid solution. The chemical composition of the solid solution then changes, approaching the equilibrium concentration. The state of supersaturation is usually achieved by rapid cooling from the single-phase region (see the binary diagrams shown in Figure 6.25). As an excess phase, a pure component, a phase, or a chemical compound containing this component may serve.

Aging is a heat treatment operation in which a supersaturated solid solution decomposes. However, aging is often referred to as the process itself of isolating an excess phase. Aging can proceed at ambient air temperature (natural aging) and during heating followed by isothermal holding (artificial aging), which provides increased diffusion mobility of atoms.

The decomposition of a supersaturated solid solution is accompanied by a change in physical and mechanical properties. An increase in strength characteristics during aging is called dispersion hardening. The dependencies of the hardness of Cu–1.68%Be bronze on holding time at different aging temperatures can be seen in Figure 6.24. At an aging temperature of 300 °C, the hardness reaches its highest value. This is because the number and size of particles of the released excess phase are optimal.

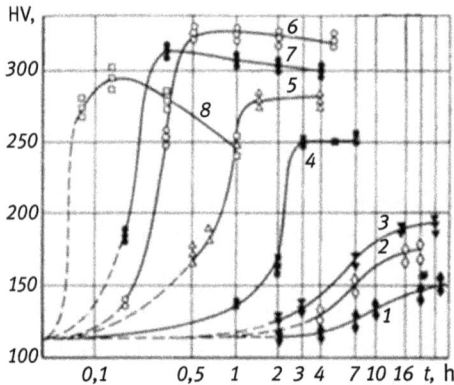

Figure 6.24: Hardness dependencies (HV) of Cu–1.68%Be on holding time at different aging temperatures, °C: 1, 100; 2, 150; 3, 200; 4, 250; 5, 280; 6, 300; 7, 330; 8, 360.

As can be inferred from the phase diagrams, the decomposition of a supersaturated solid solution takes place only when the alloy is transferred from a single-phase region into a two-phase region. Numerous examples of such diagrams are presented in Figure 6.25. The most common case is decomposition as depicted in Figure 6.25a. The diagram shows that the solubility of a second component declines with decreasing temperature. Therefore, alloys being into a single-phase state (in a certain range of concentrations) after rapid cooling (quenching) decompose upon subsequent annealing in a two-phase region (aging). This method of heat treatment is often applied for various technical alloys to obtain the desired physical and mechanical properties.

The decomposition of a supersaturated solid solution can occur by a continuous or discontinuous mechanism. The former includes the nucleation of particles and their gradual growth. A special case of the continuous decomposition is spinodal decomposition when the solid solution is stratified into phases of different chemical compositions of a similar crystalline structure (Figure 6.25f). The discontinuous

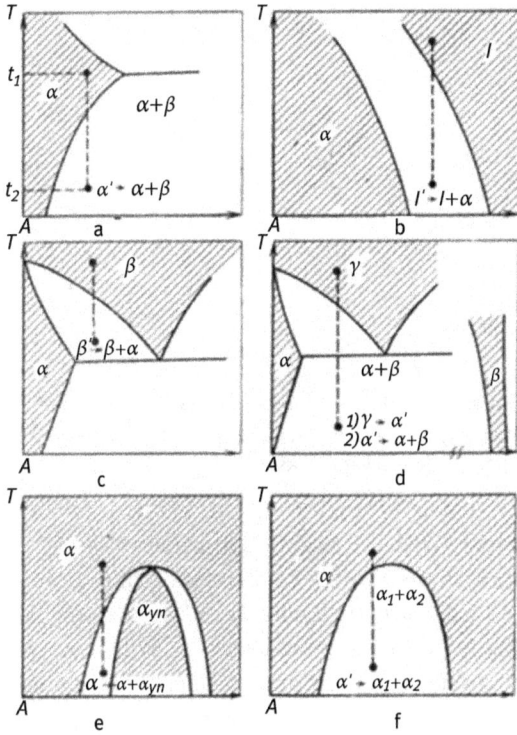

Figure 6.25: Schemes of equilibrium diagrams for alloys suffered decomposition processes.

decomposition is brought about through the growth of new sections (clusters) containing a new phase and a solid solution depleted in an excess component. Let us dwell on these types of decomposition in more detail.

6.3.1 Continuous and spinodal decomposition

Continuous decomposition is the most widespread and important for obtaining the necessary properties of industrial alloys. The thermodynamics and kinetics of the decomposition follow the type of state diagram and depend on the chemical composition of alloys and the temperature–time conditions of heat treatment. The thermodynamics of the decomposition can be illustrated by a complex hypothetical example that reflects various features of the process. Figure 6.26 presents a schematic diagram of the dependence of free energy on the chemical composition in a system decomposed on several phases (for simplicity, the temperature is fixed).

In alloys of different compositions in different states, the following phases can coexist: a supersaturated solid solution (1), metastable intermediate phases (2 and 3), and a stable (equilibrium) phase (4). In alloys having a concentration of solid solution between c_1 and c_2 at a given aging temperature, an equilibrium phase 4 with a

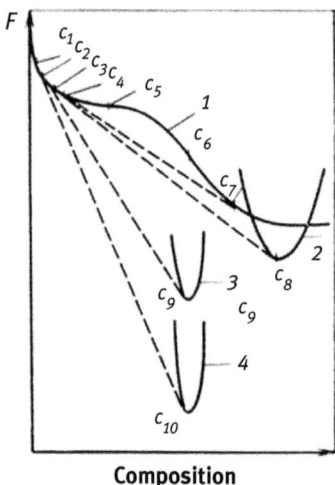

Figure 6.26: A schematic diagram of the dependence of free energy on the chemical composition in a system decomposed on several phases.

concentration c_{10} must immediately arise. With a concentration between c_2 and c_3, the decomposition is two-stage: first, a phase 3 with a concentration of c_9 occurs; the second stage involves a phase 4. With a concentration between c_3 and c_4, the first decomposition product is an ordered phase 2 with a concentration of c_8. With a concentration to the right of c_4, the first decomposition product is disordered regions (or zones) of a solid solution 1 with a concentration of c_7. In the concentration range of c_5-c_6, there is a region of spinodal decomposition.

Figures 6.27 and 6.28 illustrate examples of real diagrams with continuous and spinodal decomposition.

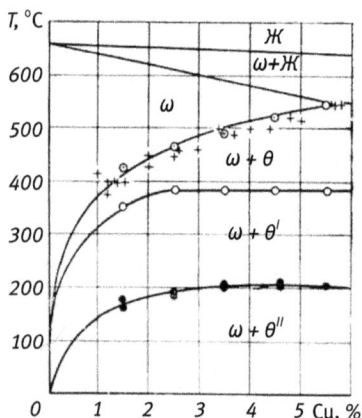

Figure 6.27: The Al angle of an Al–Cu diagram. Solid lines indicate the solubility lines for stable and metastable phases.

In the aluminum corner of the Al–Cu diagram (Figure 6.27), a solid solution of copper in aluminum ω undergoes continuous decomposition to precipitate the equilibrium

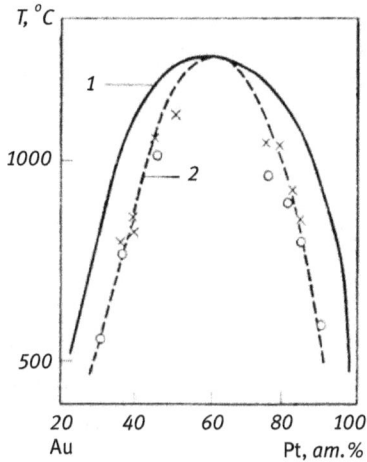

Figure 6.28: Spinodal decomposition in an Au–Pt system: 1, the delamination boundary; 2, experimental data.

θ-phase. At low aging temperatures, intermediate metastable phases θ' and θ'' release at the beginning of the decomposition. Solubility lines for all three phases are drawn in the diagram. In alloys of the Au–Pt system (Figure 6.28), decomposition occurs by the spinodal mechanism. Line 1 in the Au–Pt diagram shows the demixing border of a single-phase solid solution as does the diagram in Figure 6.25e. Line 2 indicates a spinodal plotted from the temperatures of a sharp deceleration or acceleration of the decomposition process.

The kinetics of concentration changes during continuous and spinodal decomposition is shown schematically in Figure 6.29.

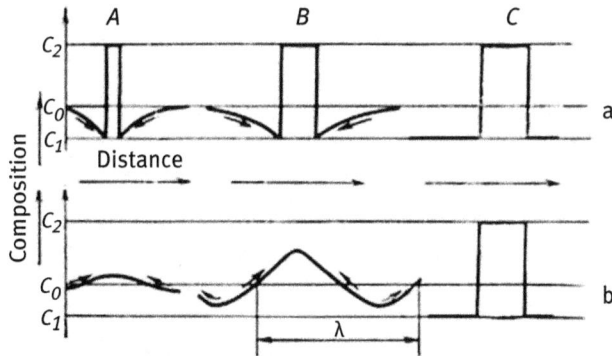

Figure 6.29: Schemes of concentration distribution at nucleation and growth (a) and during the spinodal decomposition mechanism (b). c_1, equilibrium composition of a solid solution; c_2, composition of an excess phase; A, the initial stage of decomposition; B, an intermediate stage; C, the final stage.

A solid solution changes its initial concentration c_0 when decomposed. In the process, the concentration of the solution becomes equal to c_1 to isolate an excess phase with a concentration of c_2. The equilibrium concentration distribution for these decomposition mechanisms may be the same (a final state B). However, the kinetics of the above processes is significantly different. As a result, a microstructure of different morphology is formed. During continuous decomposition, particles of the excess phase with a concentration of c_2 nucleate (state A) and grow (states B and C). The excess component diffuses from the matrix to a particle; the diffusion paths are indicated by arrows. Spinodal decomposition gives rise to sections with an increased and decreased concentration of the excess component (state A). Then, "concentration waves" with a period λ (state B) arise. This decomposition stage is responsible for so-called modulated structures. Equilibrium is achieved through uphill diffusion (according to Konobeyevsky); the diffusion paths are indicated by arrows. Thus, spinodal decomposition in a certain sense is non-nucleus.

6.3.2 Formation of microstructure

Nucleation during continuous decomposition occurs through the fluctuation emergence of regions enriched in an excess component. These regions are most often formed heterogeneously on defects in the crystal lattice (grain/subgrain boundaries, dislocations, etc.) and less often, homogeneously. The shape of the regions can be flat, spherical, or more complex. The early decomposition stage has gained the name of "zone" one. The enriched areas are called Guinier–Preston zones (GP zones), after the scientists who discovered them in Al–Cu alloys.

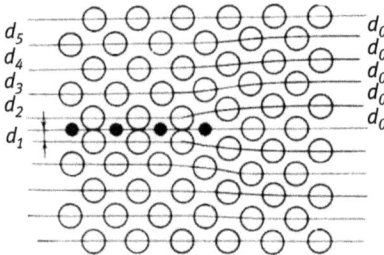

Figure 6.30: Scheme of a flat region in an Al–Cu solid solution. Clusters of Cu atoms are shown.

Figure 6.30 shows a scheme of a flat zone in an Al–Cu solid solution. Clusters of copper atoms alter interplanar distances in adjacent areas of an aluminum-based solid solution.

At the zone stage and after it, particles of a new phase are formed. Particles of intermediate, metastable phases can appear to form a stable phase, as shown in Figure 6.27 (decomposition of Al–Cu alloys). This is the rule of steps. The intermediate phase may have a lower energy barrier to emerge. This is because its chemical composition is closer to that of a solid solution rather than a stable phase. Another reason for the low energy barrier may consist in better matching the crystal lattice of the intermediate phase with the solid solution lattice than for the stable phase lattice. Consequently, the surface energy is less to create particles of the intermediate phase.

The surface energy of the interphase boundary (between the matrix and the particle) depends on the degree of its coherence. Figure 6.31 is a diagram illustrating coherent, partially coherent, and incoherent conjugation of the crystal lattices of the isolated particles and the matrix.

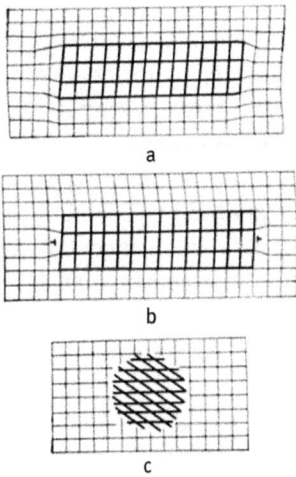

Figure 6.31: Schemes of interphase boundary structure: (a) coherent boundary, (b) partially coherent boundary, and (c) incoherent boundary.

The coherent conjugation provides the minimum surface energy of the interphase boundary. However, the mismatch between the crystal lattices of the growing particle and the matrix provokes the accumulation of elastic energy. The relaxation of the latter leads to the appearance of misfit dislocations, and the coherent interphase boundary is transformed into a partially coherent one. If the particle and the matrix differ largely in lattice parameters (according to some data, more than 15%), the interphase boundary becomes incoherent. The energy barrier in coherent nucleation being smaller than that in incoherent nucleation, orientational relations between the crystal lattices of the particle and matrix are often fulfilled to isolate phases. With accounting for the rule of steps, the decomposition process can be represented by a series of C-shaped curves (Figure 6.32).

At temperature T_3, the equilibrium phase β is immediately isolated. At T_2, the metastable phase β' is first isolated, and only then phase β appears. At T_1, the G–P zones first arise, the metastable phase β' is then precipitated followed by isolation of phase β with increasing holding time. There exists a general rule: the greater the degree of supersaturation, the more likely (with all other factors being equal) the formation of intermediate phases is.

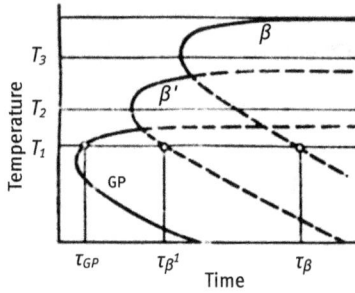

Figure 6.32: C-shaped curves of the formation of the G–P zones, intermediate (β') and stable (β) phases at aging (scheme).

The decomposition kinetics of a supersaturated solid solution is controlled, as a rule, by diffusion of the dissolved element in the matrix. Temperature and holding time determine the course of the diffusion-controlled process. For ensuring the desired microstructure and optimal service properties, the decomposition can be stopped at a necessary stage.

Let us give examples of the microstructures of alloys aged by a continuous mechanism. Figure 6.33 presents the structure of an Al–20% Zn alloy containing particles of the β' metastable aging phase (almost pure zinc is a stable phase β).

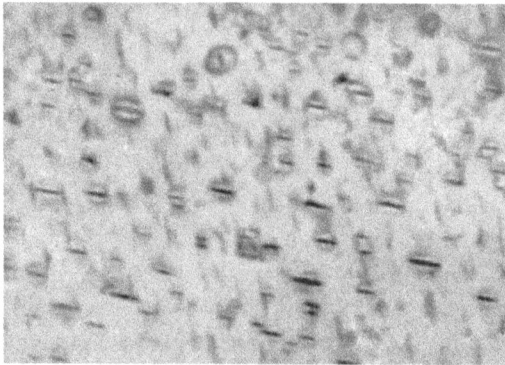

Figure 6.33: Continuous decomposition of an Al–20%Zn alloy. The contrast of elastic-distortion fields is seen around particles of β' phase, ×100,000.

Elastic-distortion fields are visible around the particles (Dobromyslov). Figure 6.34 is the image of the structure of austenitic chromium–nickel–manganese steel with carbide particles continuously precipitated.

As shown in Figure 6.35, the arrangement of particles of $Me_{23}C_6$ carbides is crystallographically ordered; it is due to their elastic interaction (Farber). The modulated structure in a Ticonal hard-magnetic alloy (Fe-Co-Ni-Al-Ti) after aging in a magnetic field is shown in Figure 6.35. Spinodal decomposition leads to the formation of a structure of alternating areas of two phases.

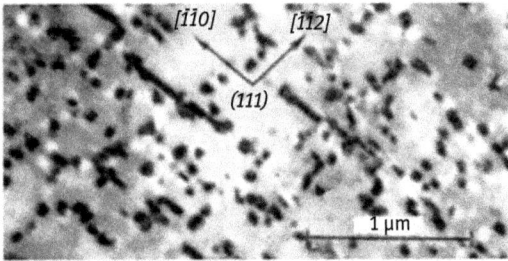

Figure 6.34: Continuous precipitation of carbides in austenite Cr–Ni–Mn steel. Row lining-up of particles $(\times 26,000)$.

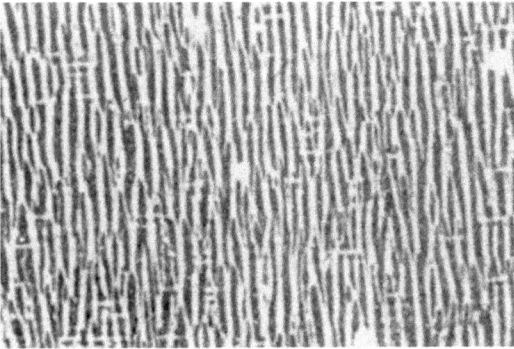

Figure 6.35: A modulated structure in a Ticonal alloy after aging in a magnetic field, × 50,000.

Separation of a new phase through a continuous mechanism can be conditionally divided into four successive stages:
- Nucleation of particles
- Particle growth until the thermodynamic equilibrium of the components at the interphase boundary
- Particle growth under equilibrium conditions at the interphase boundary, when the speed of the process is determined by diffusion or by the rate of supply (or removal) of atoms of a substance
- Particle coalescence leading to coarsening of the structure

A theoretical analysis of continuous decomposition in a binary system includes six kinetic coefficients, but namely
- Two coefficients that determine the mobility of the first component in the initial and new phases
- Two similar coefficients for atoms of the second component
- Two coefficients that determine the transition rate of atoms of each of the components through the interphase boundary

The theoretical analysis usually requires a number of assumptions made for leaving only two kinetic coefficients. The former controls supply (removal) of atoms of the solute to the growing center and is related to the *diffusion coefficient of the solute* in the solvent lattice. It is also often used for approximate separation kinetics analysis. The latter is responsible for transitioning atoms of the main components (solvent) through the interphase boundary.

6.3.3 Discontinuous decomposition

During discontinuous decomposition in grains of an initially supersaturated solid solution, areas (colonies or cells) nucleate and grow. They contain a new phase and solid solution depleted in an excess component. The precipitated phase is lamellar (less often, it has the form of rods). The colonies consist of alternating lamellae of the excess phase and depleted solid solution. As an example, Figure 6.36 shows the microstructure of a colony of discontinuous decomposition in austenitic nitrogen-containing chromium–manganese steel.

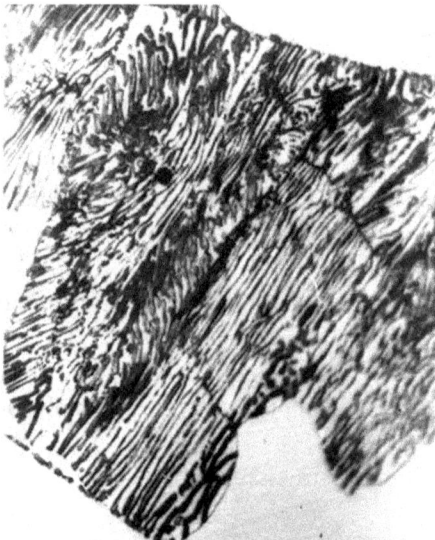

Figure 6.36: Discontinuous decomposition in austenite nitrogen-containing chromium–manganese steel, ×750.

The excess phase here is nitrogen nitride. Usually, at a fixed aging temperature, a constant distance between adjacent similarly named lamellae (an interlamellar spacing) is established. The lamella-to-lamella distance declines with decreasing the aging temperature. This pattern results from the colony growth mechanism (see below).

When decomposed, the solid solution saves its initial concentration until it disappears. At the border between the colony and the initial solution, the concentration sharply rises and reaches a concentration inside the colony. Figure 6.37 outlines concentration distribution schemes during growth upon continuous and discontinuous decomposition.

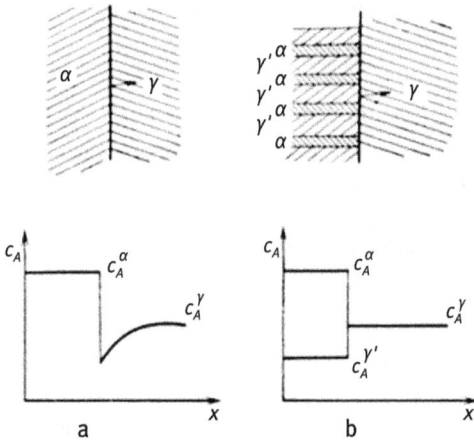

Figure 6.37: A scheme of concentration distribution under continuous (a) and discontinuous (b) decomposition.

In the first case, the concentration of the excess component near the interphase boundary gradually decreases; volume diffusion controls this process. The second case involves differentiation of the components at the colony boundary: the excess component passes from the solid solution to the aging phase along the colony boundary. The process is governed by diffusion along a moving (migrating) boundary.

A colony of discontinuous decomposition originates at a high-angle boundary and grows inward one of the adjacent grains. The orientations of the crystal lattice of the solid solution in the colony and in this grain are different. However, the orientation of the crystal lattice coincides with the orientation of the neighboring grain from which growth begins. In the process, the orientation of all lamellae of the solid solution in the colony is the same or almost the same. These facts allow one to regard the colony growth as the motion of a high-angle boundary during recrystallization, the course of which is complicated, however, by the formation of lamellae of a new phase.

The mechanism of nucleation of a colony upon discontinuous decomposition is far from being precisely ascertained. Nevertheless, there are some nucleation models based on the interaction of a migrating boundary with particles of a precipitating phase at the early stages of its onset. The growth mechanism for the colony formed is clearer. As mentioned above, the growth is controlled by diffusion along the migrating boundary. The coefficient of boundary diffusion is 3–5 orders of magnitude greater than that for volume diffusion. There are data that diffusion along the migrating boundary proceeds even faster. The diffusion paths are very short. They make up half of the interlamellar distance that is small. Therefore, discontinuous decomposition is possible to take place at low temperatures, at which continuous decomposition is almost suppressed. The boundary diffusion coefficient D_{boun}, the colony growth rate V, and the interlamellar distance S are related as follows:

$$D_{\text{boun}} = VS^2/\lambda,$$

where λ is the thickness of the boundary, usually taken as $\lambda = 0.5$ nm.

Discontinuous decomposition in commercial alloys is often undesirable, leading to low strength and ductility characteristics.

It is worth mentioning another related phase transformation called eutectoid decomposition. In this case, colony growth is controlled by diffusion along a migrating boundary as it does upon discontinuous decomposition. Eutectoid decomposition differs from the discontinuous one in that a supersaturated solid solution, when decomposed, segregates into two phases with new crystal lattices. One of the most important examples of eutectoid decomposition in practice is pearlite transformation in steels (see Chapter 7). For example, a carbon-supersaturated solid solution with a face-centered lattice (austenite) turns into a carbon-poor solid solution with a body-centered lattice (ferrite) and an excess phase – cementite (carbide with a rhombic lattice). Pearlite transformation is discussed in detail in the next chapter.

6.4 Ordering

A substitutional solid solution consisting of two sorts of atoms (A and B) of the stoichiometric composition is thought to be disordered if they randomly reside at sites of the crystal lattice. If atoms of each sort (A and B) predominantly occupy sites of different types in the crystal lattice, they are then said to have atomic ordering. For example, let a binary equiatomic composition alloy AB comprise atoms of A and B sorts at different type of sites, respectively. In such a case, the alloy is atomically ordered with a maximum degree of long-range order.

The degree of long-range order is quantitatively characterized by a long-range order parameter η introduced through the probability of substituting lattice sites of a certain type for atoms of a given type:

$$\eta = \left(p_A^{(1)} - C_A\right) \big/ (1 - v),$$

where $p_A^{(1)}$ is the probability of substituting sites of the first type for atoms of sort A. $p_A^{(1)} = N_A^{(1)}/N^{(1)}$; $N_A^{(1)}$ is the number of atoms of sort A at sites of the first type; $N^{(1)}$ is the number of sites of the first type; C_A is the concentration of atoms of sort A, $C_A = N_A/N$, N_A is the number of atoms of sort A, N is the total number of atoms; v is the relative concentration of sites of the first sort, $v = N^{(1)}/N$. v is determined by the structure of the crystal lattice. The long-range order parameter varies within $\eta_{max} = 1$, $\eta_{min} = 0$.

An increase in temperature disturbs the long-range order in a fully ordered alloy. The latter becomes disordered upon reaching a certain characteristic temperature, and the long-range order disappears. The temperature T_C of disordering (ordering) is called the Curie ordering temperature or the Kurnakov point (earlier definition). The disappearance of the long-range order can occur either abruptly or gradually.

Ordering can take place as a first-order phase transition (e.g., CuAu alloy) or as a second-order phase transition (e.g., β-brass). In the former, the degree of long-range order η at an ordering temperature T_0 changes abruptly. The first derivatives of the thermodynamic potential also have a jump: heat is absorbed (liberated), and entropy and specific volume dramatically rise. The transformation occurs by nucleation and growth of areas of an ordered phase in a disordered matrix. The chemical compositions of the ordered and disordered phases may differ. The phase diagram should include a two-phase region.

A continuous increase in the degree of long-range order η from zero to the maximum value with decreasing temperature below T_0 stands for a second-order phase transition. In this case, there is no heat of transition; the change in energy and specific volume is equal to zero. Nevertheless, the second derivatives of the thermodynamic potential as specific heat and the thermal-expansion coefficient have a jump. The chemical composition of the ordered and disordered phases is the same. The phase transition may hold without the appearance of a two-phase region.

Let us give a few examples of crystalline structures of ordered alloys (they are also called superstructures).

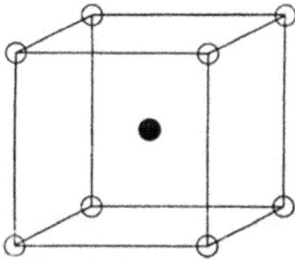

Figure 6.38: The unit cell of an ordered alloy with a structure of the β-brass type (CsCl or B2).

Figure 6.38 shows the simplest structure of the β-brass type (CsCl). Atoms of one sort (e.g., sort A) are located at the vertices of a cubic body-centered cell, and atoms of another sort (B) reside in the center of the cell. CuZn, CuPd, FeAl, and FeCo alloys have such a structure near the equiatomic composition. Ordering in these alloys occurs as a first-order phase transition. The arrangement of atoms in the superstructure of Cu_3Au (structure of type $L1_2$) can be viewed in Figure 6.39.

Atoms of gold are located at the vertices of a cubic face-centered cell; copper atoms are located in the centers of faces. With increasing temperature, the degree of long-range order of this alloy first declines to a certain value and then drops abruptly to zero at the ordering temperature T_C. Ordering in this alloy occurs as a first-order phase transition. Figure 6.40 sketches the unit cell of an ordered CuAu alloy ($L1_0$-type structure).

Atoms of gold occupy the vertices of a cubic face-centered cell, and centers of two opposite faces and atoms of copper are arranged in the centers of the other four

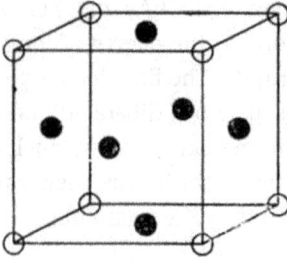

Figure 6.39: The unit cell of an ordered alloy with a Cu_3Au (LI_2) structure.

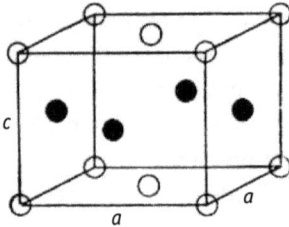

Figure 6.40: The unit cell of an ordered alloy with a CuAu ($L1_0$) structure.

faces. It is seen that the (001) planes are alternately filled either with gold atoms or with copper atoms. As a consequence, a layered structure arises, with its cubic symmetry being not preserved, and the crystal lattice becomes tetragonal. CoPt, FePt, and FePd alloys have such a structure. In these alloys, ordering occurs as a first-order phase transition.

The transition from a disordered state to an ordered state is carried out by diffusion of atoms over distances of the order of interatomic ones. A low diffusion rate at low temperatures makes it possible to obtain and maintain a disordered metastable state. Among the methods to achieve this goal are quenching from a temperature exceeding the Curie ordering temperature or disordering by cold deformation. The foregoing allows studying the kinetics of the ordering process and the formation of the structure of an ordered alloy under isothermal holdings.

Direct methods for investigating ordering are radiography, neutron diffraction, and transmission electron microscopy. If an AB alloy is in a disordered state, all parallel planes of the crystal lattice have the same average composition and are equivalent to the scattering of X-rays (and neutrons). The X-ray or neutron diffraction pattern contains structural reflections. If the alloy becomes ordered, parallel planes can have different average composition and are nonequivalent to scattering. Then, the X-ray diffraction pattern (neutron diffraction pattern) may include, apart from structural reflections, additional reflections called superstructural. Their intensity depends on the difference in the concentrations of atoms of any sort in these planes and, therefore, on the degree of long-range order. The superstructural-reflection intensity I has the form

$$I = C \left| f_A - f_B \right|^2 \eta^2,$$

where f_A and f_B are atomic factors for the scattering of X-rays by atoms of A and B sorts, and C is the proportionality coefficient. This relation enables one to determine the long-range order parameter from diffraction measurements.

Being small, the difference in the atomic factors of X-ray scattering by atoms of different sorts causes no or low intensity of superstructural reflections. In this case, a neutron diffraction study may be useful. In neutron diffraction, the neutron scattering amplitude plays the same role as so does the atomic scattering factor in X-radiography. So, the atomic scattering factors of iron and cobalt in an ordered FeCo alloy are close, and the intensities of superstructural reflections are difficult to detect. At the same time, the neutron scattering amplitudes are very different, and superstructural reflections are of high intensity. The neutron scattering amplitude can be either positive or negative. If the amplitudes of neutron scattering by atoms of different sorts have a different sign, then the intensity of superstructural reflections is greater than that of structural ones. As an example, we give an ordered NiTi alloy, in which the neutron scattering amplitudes of nickel and titanium atoms have different signs. Another advantage of the neutron diffraction method as compared to the X-ray method is that the amplitude of neutron scattering by an atom (nucleus) is scattering angle independent. This fact appreciably simplifies the diffraction measurement technique and result processing.

Transmission electron microscopy is the most informative method for studying ordering, because it allows, along with the study of the crystal structure by diffraction methods, one to observe the formation of the microstructure. An example of applying this method to the study of ordering in a CuAu alloy is given below.

Ordering of alloys significantly impacts on their physical and mechanical properties. A classic example is the influence of ordering on electrical resistance in a Cu–Au alloy system. Figure 6.41 demonstrates the dependencies of the electrical resistance on the chemical composition of the alloys in a disordered state (curve a) and after ordering (broken curve b).

At concentrations corresponding to CuAu and Cu_3Au, the curve b meets minimums of electrical resistance. The dashed line c refers to the concentration dependence of the electrical resistance, additively composed of the copper and gold resistances. Historically, minima of electrical resistance in a Cu–Au system were discovered before ordering physics arose. N.S. Kurnakov, S.F. Zhemchuzhny, and M.M. Zasedalatev were the first to fix the minima for the compositions of CuAu and Cu_3Au in their research works. Once discovered, this phenomenon became the basis for the development of the physics of ordered alloys. It is this fundamental finding that enables G. Tamman to put forward the hypothesis of existing ordered structures.

The formation of a structural state with high values of coercive force and magnetic energy in magnetically hard CoPt, FePt, and FePd alloys, when ordering, constitutes a striking example of the influence of atomic ordering on magnetic properties. Another example is the negative influence of atomic ordering on the magnetic permeability in iron–nickel alloys of the Permalloy type. These disordered alloys with

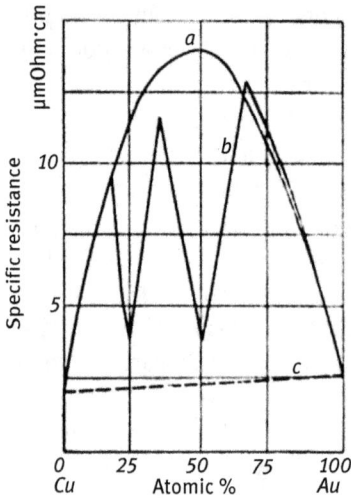

Figure 6.41: Electrical resistance of alloys in a Cu–Au system in the disordered (a) and ordered (b) states.

70–80%Ni turn into ordered ones when slowly cooled to room temperature. However, their disordered state is retained upon rapid cooling (quenching). In a disordered state, the magnetic permeability is greater than that after ordering.

Ordering in CuAu alloys greatly affects the mechanical properties. The strength characteristics of the alloy increase several times upon achieving the so-called small-domain state. In the process of ordering, a complex microstructure called a domain structure is formed in each grain of the disordered phase (do not confuse it with a domain structure in ferromagnets). Domains in an ordered alloy have a different nature. First, these are domains of thermal origin. This is because the areas of the ordered phase in each grain grow from differently oriented nuclei that have arisen independently. The grown-up areas adjoin to each other along the boundaries, which are called antiphase boundaries of ordering. The simplest two-dimensional diagram of the antiphase boundary is presented in Figure 6.42.

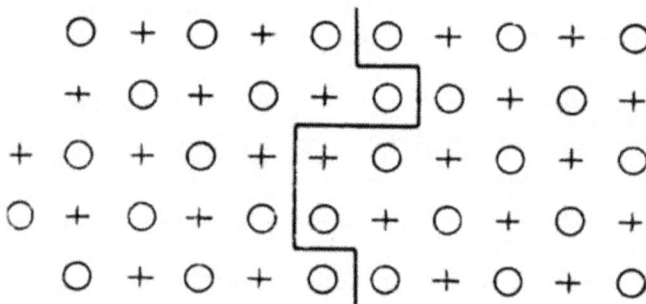

Figure 6.42: A two-dimensional scheme of the antiphase boundary.

Along the boundary, atoms of the same sort are adjacent, that is to say, the order is disturbed. Images of the antiphase boundaries of ordering can be visualized by using dark-field electron microscopy. Figure 6.43 exemplifies the antiphase boundaries in a $Cu_3Au + 4\%$ Ag alloy.

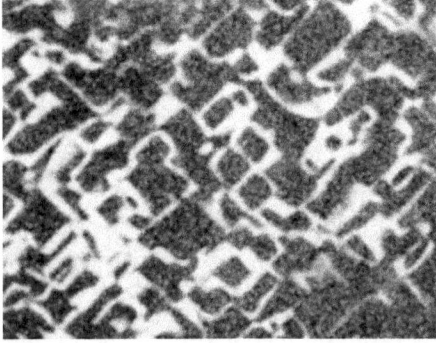

Figure 6.43: A dark-field image of the antiphase boundaries in an ordered $Cu_3Au + 4\%Ag$ alloy($\times 100,000$).

Second, there arise domains whose nature is associated with the crystal-structural ordering mechanism. The tetragonality axis c of the ordered phase $L1_0$ may come from each of the three axes of the <100> type of the disordered phase. Therefore, three types of domains (c-domains) are possible to form. The c axes in these domains are misoriented by 90°, without taking the lattice tetragonality into account. In reality, the tetragonality of the ordered phase lattice and the transformation mechanism generate 12 types of domains (Syutkina). Domains join along {110} planes. In each of the six {110} planes, two full-coherently conjugated domains are formed. Their alternation minimizes the arising elastic stresses and forms an area of the ordered phase. In an ordered state, the microstructure consists of such areas (Figure 6.44).

Figure 6.44: A domain structure in a CuAu alloy produced at high-temperature ordering ($\times 50,000$).

Such a structural mechanism of ordering is characteristic of "high-temperature ordering" upon cooling from a disordered state, at a relatively low nucleation rate. The disordered state fixed by quenching followed by ordering at a low heating temperature ("low-temperature ordering") gives rise to a dispersed small-domain structure due to a high nucleation rate (Figure 6.45a).

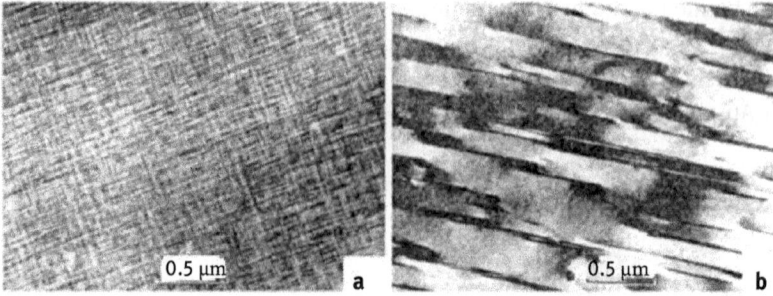

Figure 6.45: A fine-domain structure in a CuAu alloy produced at low-temperature ordering (a) and domain coarsening at additional heating (b) ($\times 50,000$).

This state is characterized by a high level of elastic stresses, and domain growth occurs upon heating (Figure 6.45b).

In conclusion of this section, we briefly dwell on the concept of short-range order. In the case of a binary alloy, the lattice in a fully ordered state consists of correctly alternating atoms of two sorts (A and B). Otherwise speaking, the nearest neighbors of each atom are atoms of another sort. A non-full order makes itself felt in appearing atoms of the same sort among the neighbors. Let p be the average number of nearest neighbors of another sort, and q be the average number of neighbors of the same sort. Then, the degree of short-range order can be called the following quantity:

$$\xi = (p - q)/(p + q).$$

As an example, we give the short-range order in Fe–Ni and Fe–Ni–Co alloys upon annealing in a magnetic field. The peculiar form of short-range order (directional order) contributes to a rectangular-shaped magnetic hysteresis loop.

Chapter 7
Transformations in steel

Steels are iron–carbon alloys (carbon steels), and the same alloys additionally alloyed with other elements (alloyed steels). Steels can undergo multiple phase transformations due, firstly, to polymorphism of iron, and secondly, the precipitation and dissolution of carbides. Analysis of phase transformations in steels requires constructing diagrams to describe their thermodynamics and kinetics. Among these are iron–cementite–alloying element phase diagrams and decomposition diagrams of supercooled austenite. The former are responsible for thermodynamics of the transformations; the latter reflect their kinetics. The iron–cementite diagram is metastable. However, this diagram can be accepted as conditionally stable to analyze the overwhelming majority of transformations in steels. This is because equilibrium in an iron–graphite system can be reached only during extremely long holding. The same applies to the ternary iron–cementite–alloying element diagrams.

Such a complex system as steel forms numerous metastable (intermediate) phases and structural states. To describe the temperature–time conditions for similar states to emerge, decomposition diagrams of supercooled austenite are plotted, which delineate the kinetics of transformations.

7.1 Iron–cementite diagram

Figure 7.1 shows a phase diagram of iron–cementite (Fe_3C).

The dashed lines indicate the equilibrium iron–graphite diagram. Iron polymorphism is manifested as follows. Iron crystallizes in the BCC modification (δ-ferrite). When cooled, δ-ferrite turns into the FCC modification (γ-phase, or austenite). At a temperature of 911 °C, austenite passes back into the BCC phase or the α-phase (ferrite). At this temperature, it is paramagnetic. Below 768 °C (Curie point), it becomes ferromagnetic. Previously, paramagnetic austenite was designated as the β-phase. The iron–cementite diagram reproduces three isothermal reactions such as peritectic at 1,493 °C, eutectic at 1,147 °C, and a eutectoid at 727 °C. The maximum solubility of carbon in austenite (2.1 wt% C, point E) is the ultimate composition separating steel and cast iron. The eutectoid point (0.80 wt% C, point S) splits steels into pre-eutectoid and hyper-eutectoid ones. Below the eutectoid temperature, there is a two-phase region in which ferrite and cementite are in equilibrium. The solubility of carbon in ferrite is substantially less than that in austenite. At the eutectoid temperature, it amounts to only 0.0206 wt%, while at 500 °C it is less than 0.001%. The solubility of carbon in austenite and ferrite varies with temperature, and the aging process is possible to proceed in steels. As will be clear from what follows, the eutectoid reaction is important for microstructures in steels to form.

https://doi.org/10.1515/9783110758023-007

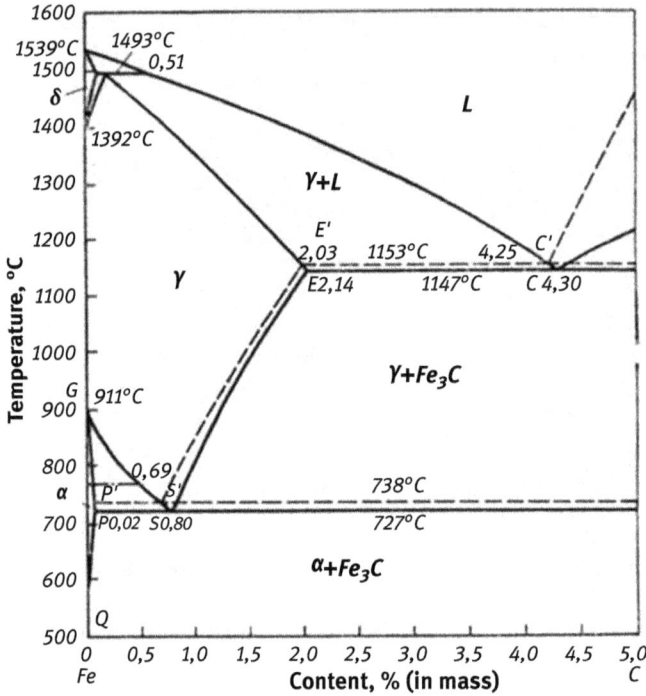

Figure 7.1: An iron–cementite state diagram.

7.2 Diagrams of the decomposition of supercooled austenite (TTT diagrams)

Figure 7.2 presents a decomposition diagram of supercooled austenite for pre-eutectoid steels.

Constructing such diagrams requires performing a few steps. The first step involves rapid transfer of samples from the temperature region of stable austenite to the region of supercooled austenite. The latter contains the decomposition process during isothermal holding at various temperatures. The temperatures of the decomposition onset initiate the so-called C-shaped curves in time–temperature coordinates. Similar curves can be constructed for both the ends of decomposition and any degree of transformation. The C-shape of the curves shows that the incubation period of decomposition is the least value in a certain "average" temperature range. An explanation of this fact will be given in Section 7.4.

The diagrams of supercooled-austenite decomposition for alloyed steels can have a more complicated shape and consist of two C-shaped curves (Figure 7.3).

In such cases, the high-temperature part of the diagram corresponds to pearlite transformation (see below), and the low-temperature part corresponds to bainitic

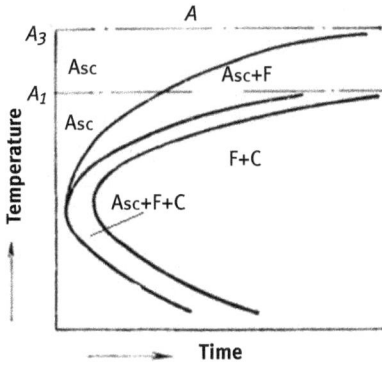

Figure 7.2: A diagram of isothermal decomposition of austenite in pre-eutectoid steel: A, stable austenite; A_{sc}, supercooled austenite; F, ferrite; C, carbide; A_1 and A_3, temperatures of the start and finish of the formation of austenite when heating.

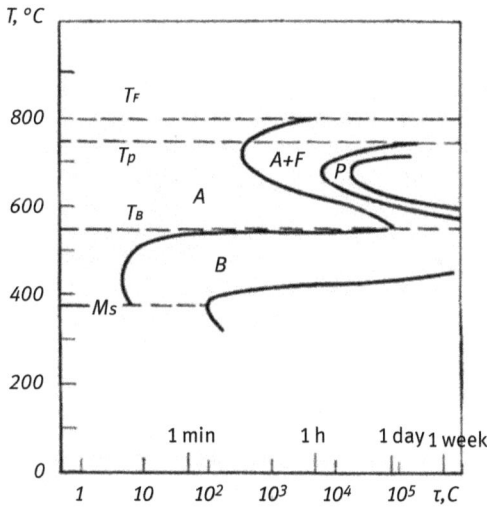

Figure 7.3: A diagram of isothermal decomposition of supercooled austenite in alloyed steel: A, austenite; B, bainite; T_F, precipitation of ferrite; T_P, formation of pearlite; T_B, formation of bainite; M_s, martensitic point.

transformation (see below). The decomposition of austenite can begin with the release of "excess" phases such as ferrite (Figure 7.2) and cementite in pre-eutectoid and hypereutectoid steels, respectively. With increasing holding time, a pearlite reaction develops.

In the diagram of decomposition of austenite, a cooling line may be drawn in the time–temperature coordinates, from the region of stable austenite. If the cooling rate is greater than some critical one (the cooling line lies to the left of the C-shaped curve at the onset of decomposition), austenite does not decompose when cooled. At low temperatures, the supercooled austenite turns into martensite. In other words, the quenching of martensite occurs. If the temperature of the martensitic transformation onset lies below the temperature of the quenching medium, then austenite remains unconverted. If the cooling rate is less than a critical one, the cooling line crosses the C-curve, and pearlite or bainitic transformation begins.

7.3 Main transformations in steels

It follows from the previous section that the main transformations in steel during cooling are pearlitic, martensitic, and bainitic (intermediate) transformations.

The main transformations during heating are tempering of martensite and the formation of austenite. More precisely, the tempering of martensite is a technological operation during which an unstable structural state that arises during martensitic transformation approaches equilibrium. Tempering of martensite consists of several transformations taking place in different temperature ranges. With an increase in the tempering temperature, the state of the martensite crystal lattice changes, carbides precipitate, and alloying elements and impurities are redistributed between the bulk and near-grain zones of grains. Such an important and, in turn, undesirable phenomenon as tempering brittleness is associated with complex processes during tempering (see below).

The formation of austenite should be considered for two different initial structural states: ferrite + cementite and martensite. Figure 7.4 outlines a kinetics diagram of the isothermal formation of austenite in pre-eutectoid steel with the initial ferrite–pearlite structure.

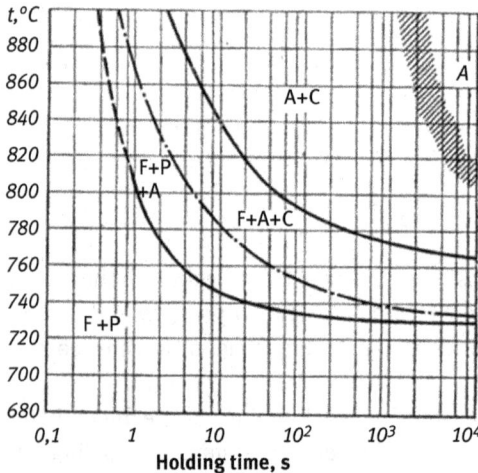

Figure 7.4: A diagram of the kinetics of the isothermal formation of austenite in pre-eutectoid steel with an initial ferrite–pearlite structure.

The lines in the diagram stand for the beginning of the formation of austenite, the finish of the turning of pearlite into austenite, and the complete dissolution of carbides. As can be seen from the figure, equilibrium is reached in a more complex way than the state diagram dictates. With an initial martensitic structure, the carbide phase does not complicate the formation of austenite. However, under certain conditions, the latter is formed, involving the phenomenon of structural heredity (see below).

7.4 Pearlite transformation

Pearlite transformation occurs in the two-phase region of ferrite + cementite state diagrams of iron–cementite. Pearlite is referred to as a two-phased microstructure composed, as a rule, of alternating lamellae of ferrite and cementite (Figure 7.5).

Figure 7.5: Thin-lamellar pearlite in eutectoid steel, × 40,000.

Such lamellar structures are usually a result of eutectoid reaction or discontinuous decomposition. The morphology of the lamellar structure is explained by the mechanism of its growth, including diffusion along the migrating boundary (see below).

The transformation kinetics is governed, as usual, by the rate of nucleation of centers and the rate of their growth. Let us look into the change in the nucleation rate due to supercooling below the transformation temperature (eutectoid line temperature). With falling temperature, the magnitude of supercooling ΔT increases, and the work A for a critical size nucleus to emerge declines. The nucleation rate of center N rises exponentially:

$$N = K_1 \exp(-A/kT),$$

where k is the Boltzmann constant and K_1 is a coefficient accounting for the rate of transition of atoms to the nucleus (the fluctuation displacement rate similar to the diffusion rate). However, with lowering temperature, the mobility of atoms drops.

Therefore, the foregoing should be taken into account by replacing the coefficient K_1 by the factor $K_2\exp(-U/kT)$, where U is the activation energy of the fluctuation transition of atoms to the nucleus. Thus, the expression for the nucleation rate consists of two exponentials:

$$N = K_2\exp(-A/kT)\exp(-U/kT).$$

Upon small supercoolings, the nucleation rate is determined by the work A spent on the creation of a nucleus. However, at large supercoolings, it is determined by the activation energy U of atomic displacements. Two mutually opposite factors control the maximum position in the nucleation rate–supercooling curve (Figure 7.6).

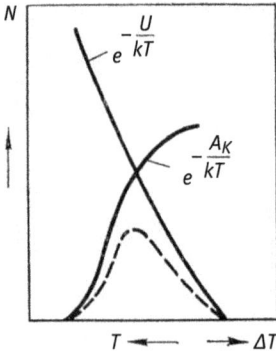

Figure 7.6: Nucleation rate dependence of center N on supercooling ΔT(dashed line).

A similar dependence exists for the growth rate but the maximum positions of the growth and nucleation rates may not coincide. The maximum values of the growth and nucleation rates yield C-shaped decomposition diagrams for supercooled austenite, with the incubation period becoming minimal in the "average" temperature range.

L.I. Mirkin derived the following formula for the kinetics of transformation:

$$V_t/V_0 = 1 - \exp\left(\tfrac{1}{3}\pi N_V G^3 t^4\right),$$

where V_t/V_0 is the fraction of the volume exposed to transformation during a time t, N_V is the nucleation rate in the volume, and G is the growth rate. The formula implies that a change in the growth rate significantly affects the overall transformation rate than a change in the nucleation rate.

Minimization of the surface energy between the regions of the initial and growing phases is achieved by the similarity in the arrangement of atoms at interphase boundaries. This leads to the fact that certain orientation relations hold between the phases involved in the transformation. Between pearlite ferrite and austenite, the Kurdjumov–Sachs orientation relationships are fulfilled. In this case, the arrangement of atoms in adjacent closely packed parallel planes has the maximum similarity:

$$(011)_F \| (111)_A \quad \text{and} \quad [1\bar{1}1]_F \| [10\bar{1}]_A.$$

Between the lattices of cementite and austenite, two types of orientation relations were observed:

$$[100]_C \| \langle 5\bar{4}5 \rangle_A, \quad [010]_F \| \langle 10\bar{1} \rangle_A, \quad [001]_F \| \langle 252 \rangle_A;$$

$$[100]_C \sim \| \langle 3\bar{1}1 \rangle_A, \quad [010]_C \sim \| \langle 3-2-9 \rangle_A, \quad [001]_C \sim \| \langle 39\bar{1} \rangle_A.$$

The former can be regarded as the Pitsch–Arbuzov–Kurdjumov orientation relationships; the latter is the Pitsch–Kutelia orientation relationships. These relations were observed in steels of different chemical compositions after transformations under various conditions. Between the lattices of cementite and ferrite, two types of orientation relations were also found:

$$(103)_C \| (101)_F, \quad (110)_C \| (111)_F.$$

$$[100]_C - 2.6° \| \langle 3\bar{1}1 \rangle_F, \quad [010]_C - 2.6° \| \langle 131 \rangle_F, \quad [001]_C \| \langle \overline{21}5 \rangle_F.$$

The former is the Bagaryatsky–Isaichev orientation relationship, and the latter is the Petch–Sukhomlin–Kutelia relationship. If the Pitsch–Arbuzov–Kurdjumov relation is fulfilled between cementite and austenite, cementite and ferrite are related by the Bagaryatsky–Isaichev relationships. If the Pitsch–Kutelia orientation relationships are established between cementite and austenite, cementite and ferrite are related by the Petch–Sukhomlin–Kutelia orientation relationships. A detailed analysis of the orientation relationship of phases during pearlitic transformation can be found in the specialized literature. However, the structural components of pearlite (ferrite and cementite) often do not have an orientation connection with the initial austenite, since pearlite originated in another austenitic grain or at the inclusion boundary.

The transformation of austenite to pearlite is a complex process. It includes the following processes: the repackaging of atoms during the polymorphic austenite–ferrite transformation, the redistribution of carbon, the formation of a cementite lattice, as well as the redistribution of alloying elements between phases in alloyed steels exposed to transformation, possibly, the formation of special carbides. The repacking rate of atoms is governed by the self-diffusion of iron. The redistribution rate of carbon between ferrite and cementite is controlled by the diffusion mobility of carbon. When analyzing the nucleation and growth of a single site of pearlite (a pearlite colony), a process that limits the rate of transformation should be identified. This process is usually carbon diffusion. In the early stages of transformation, the rate of transformation is determined by the rate of nucleation. The predominant sites of nucleation of centers are the austenite grain boundaries and other surfaces. Therefore, the nucleation rate is a structurally sensitive characteristic. The growth rate, on the contrary, is a structurally insensitive characteristic; it is constant in time at a given transformation temperature.

The end-face and lateral mechanisms of pearlite colony growth are usually considered. The former consists in carbon-depleting the austenite sites adjacent to the resulting cementite layer. In these sites, a ferrite layer is formed next to the cementite layer. Pushing from the side surface of ferrite to austenite, carbon creates conditions for another cementite layer to form. The process is repeated many times and leads to the formation of a set of alternating parallel cementite–ferrite lamellae. According to the iron–cementite diagram, pearlite in steels of the eutectoid composition contains 88% ferrite and 12% cementite. The kinetics of lateral growth is limited by the rate of bulk diffusion of carbon. The theory of the end-face growth mechanism was put forward by Zener and Hillert and supplemented by Turnbull. This mechanism is similar to the discontinuous decomposition colony growth mechanism. Ahead of the colony front, a concentration gradient is observed; carbon atoms travel toward the growing cementite layer. Diffusion can proceed both throughout the austenite volume and along the moving austenite/ferrite boundary. The diffusion paths are connected with the ferrite layer thickness and are equal to half the thickness. Figure 7.7 presents the scheme of the colony site and the direction of the diffusion fluxes of carbon.

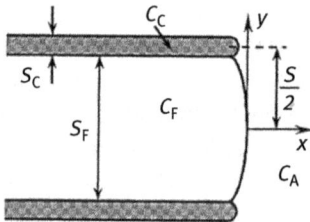

Figure 7.7: Scheme of the end-face growth of a pearlite colony: x, growth direction; S, interlamellar spacing; c_A, c_F, and c_C, carbon concentrations in austenite, ferrite, and carbide, respectively.

The rate of boundary diffusion of carbon is several orders of magnitude greater than that of bulk diffusion. Accounting for boundary diffusion allows one to better align the calculated colony growth rate and experimental data. A detailed analysis of the end-face growth mechanism of a pearlite colony can be found in the specialized literature.

An important characteristic of pearlite is the interlamellar distance, that is, the distance between the midpoints of adjacent identical lamellae. The smaller this distance, the shorter the paths for carbon atoms to diffuse, and the faster the carbon redistribution occurs for growing a colony. However, as the interlamellar distance decreases (or the ferrite and cementite lamellae get thinner), the overall area of the phase interface surfaces increases. Consequently, the surface energy rises too. If one follows the requirement of a maximum colony growth rate (Zener principle), an interlamellar distance that best satisfies these two opposite conditions is established in the process of colony growth. For each transformation temperature (and the degree of supercooling), there is an "optimal" value of the interlamellar distance. At low transformation temperatures, the diffusion rate diminishes, diffusion paths become shorter, and the interlamellar distance decreases. Figure 7.8 outlines the dependence of the interlamellar distance on the supercooling temperature.

S,cm
10^{-4}

10^{-5}

10^{-6}

700   600 500 400 t,°C

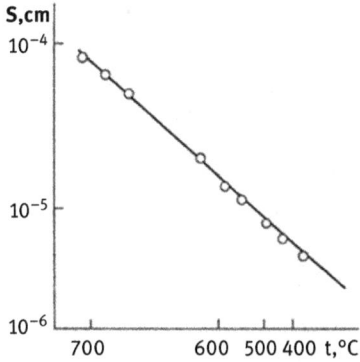

Figure 7.8: Temperature–transformation dependence of the interlamellar spacing in pearlite.

The lamella-to-lamella spacing varies from 1 μm at high transformation temperatures to 0.1–0.2 μm at low ones. Pearlite is called coarse-lamellar or fine-lamellar structures. The mechanical properties of pearlite-structure steels substantially depend on the magnitude of interlamellar spacings. Thus, the tensile strength (σ_v) of fine-lamellar pearlite formed in steel 70 at 480 °C is equal to 1,180 MPa, and the strength of coarse-lamellar pearlite formed at 600 °C amounts only to 650 MPa.

In conclusion, it should be underscored that the theory of pearlite formation for carbon steels is mainly constructed. However, as for alloyed steels, this problem becomes too involved due to the need to account for both the influence of alloying elements on the diffusion mobility of atoms and the formation of special carbides in pearlite (instead of cementite).

7.5 Martensitic transformation

The general laws and the phenomenological (crystallographic) theory of martensitic transformations are set forth in the previous chapter. Here, we will dwell on martensitic transformations in iron and steel. However, a partial repetition of the general statements cannot be dispensed with.

Against pearlite transformation, the main characteristics of the martensitic transformation are as follows.

1. A characteristic pattern (protrusions and depressions) is formed on a flat polished surface, indicating a shear transformation mechanism.
2. There occurs no decomposition of supercooled austenite, and the carbon concentration in martensite is the same as in the initial austenite. The martensitic transformation is diffusionless.
3. The martensite crystal lattice is oriented naturally relative to the austenite lattice. The orientation connection of the lattices is more stringent than during pearlite transformation.

4. Martensitic crystals are in the shape of lamellae growing at a high speed (of the order of 1 km/s).
5. The martensitic transformation has no incubation period and begins at an M_s temperature that is cooling rate independent in a wide rate range usually used. The M_s temperature dependence plotted for extremely high cooling rates can be found in the special literature (see below). To develop the transformation, it is necessary to continuously lower the temperature in the range of martensitic transformation. Once the transformation finishes, a certain amount of retained austenite is usually left.

The latter characteristic, as will be clear from what follows, is not unconditional.

Martensite is a metastable phase, and it is absent in the iron–cementite diagram. In the entire temperature range, the free energy of martensite is greater than that of the ferrite + cementite mixture. However, the free energy of martensite becomes less than that of supercooled austenite when cooling below the temperature T_o (Figure 7.9).

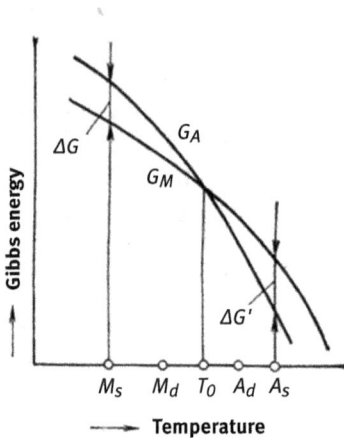

Figure 7.9: Temperature dependence of free energies (Gibbs energies) of austenite (G_A) and martensite (G_M): T_0, equilibrium temperature; M_s and M_d, temperatures of the start of martensitic transformation upon cooling and under deformation; A_s and A_d, temperatures of martensite–austenite transformation at heating and under deformation.

Any further supercooling (at a temperature of M_s) turns austenite into martensite. The T_o line of equality of the free energies of martensite and austenite and the position of the martensitic points M_s in the iron–cementite system are demonstrated in Figure 7.10.

The martensitic transformation cannot begin at the temperature T_0 since the appearance of a new phase is associated with an increase in the elastic deformation energy E_{elas} due to the shape deformation and the bulk effect of the transformation and the surface (interphase) energy E_{surf}. To compensate for the increase in elastic and surface energies, supercooling below the T_o temperature is required. In the iron–cementite system, the $T_o - M_s$ temperature difference is 200–250 K (Figure 7.10).

The kinetics of martensitic transformation in steels and iron alloys can be athermal and isothermal. With athermal kinetics, martensite is formed when cooled.

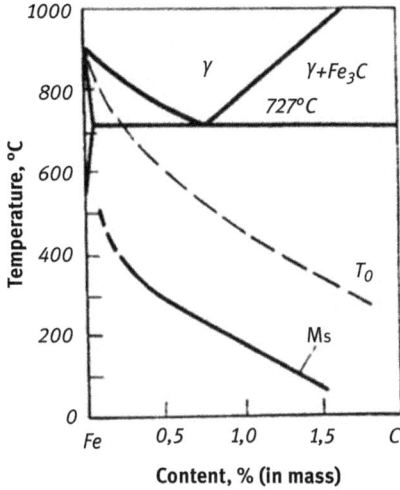

Figure 7.10: A part of the iron–cementite diagram with T_0 and M_s lines: T_0, equality of temperatures for free energies of austenite and martensite of the identical composition.

With isothermal holding, the transformation almost does not continue. Athermal transformation can be "explosive" in nature, in which up to 50% of martensite is formed in fractions of a millisecond. Martensite is formed as groups of crystals rather than as separate crystals. The growth of one crystal in a group initiates the formation of a neighboring one. In this, autocatalytic transformation consists. Such transformation usually takes place at low temperatures and is accompanied by a sound effect. With isothermal kinetics, the amount of martensite increases at isothermal holding. Figure 7.11 shows a schematic temperature dependence of the amount of martensite formed upon cooling for athermal and explosive kinetics and the dependence on the holding time for isothermal kinetics.

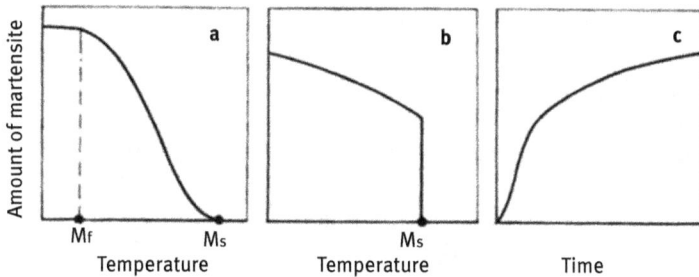

Figure 7.11: Kinetics of athermal (a), explosive (b), and isothermal (c) martensitic transformations.

As can be seen from experimental data in Figure 7.12, the isothermal kinetics of the martensitic transformation can be represented in the form of C-shaped curves in time–temperature coordinates. (In this case, the position of the martensitic point depends on the cooling rate.)

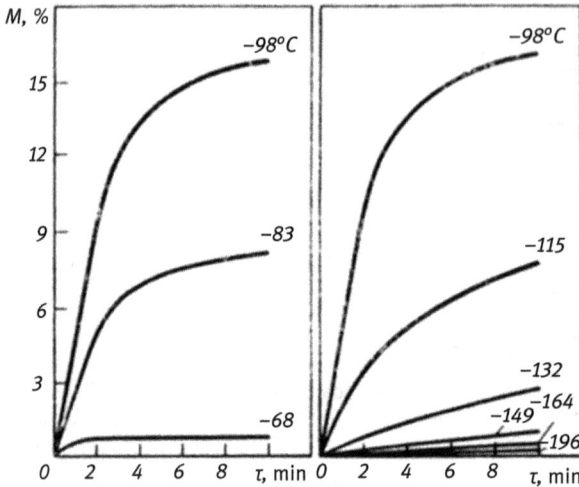

Figure 7.12: Isothermal martensitic transformation in alloyed steel during stoppages upon cooling.

The discovery of isothermal kinetics was an important step in understanding the nature of the martensitic transformation. Kurdjumov came to the conclusion that the martensitic transformation, like other transformations in the solid state, occurs through nucleation and growth. However, the problem of nucleation during martensitic transformation has not yet been solved.

The martensite crystal lattice in carbon steels is tetragonal. The degree of tetragonality linearly depends on the carbon concentration:

$$c/a = 1 + 0.046\,(\%\,C).$$

Figure 7.13 shows the experimental dependencies of periods c and a on the carbon concentration. Period c increases with increasing carbon concentration; period a diminishes slightly.

The appearance of tetragonality is explained by the ordered arrangement of carbon atoms throughout the interstices of the martensitic crystal lattice. Martensite in carbon steels is a supersaturated solid solution of carbon incorporation into α-iron. Carbon atoms occupy the octahedral pores in the BCC lattice (the centers of the faces and the middle of the edges of the unit cells) (Figure 7.14). An arrangement of carbon atoms along one of the directions of the $<001>$ type, which becomes the c-axis (in Figure 7.14 this direction is [001]), is energetically favorable.

The microstructures formed during martensitic transformations in iron alloys are very diverse. To date, more or less detailed classifications of martensite according to various criteria have been proposed. The most convenient and common classification is the classification according to morphology. In other words, martensitic crystals are classified according to the shape and ways of grouping. The shape of

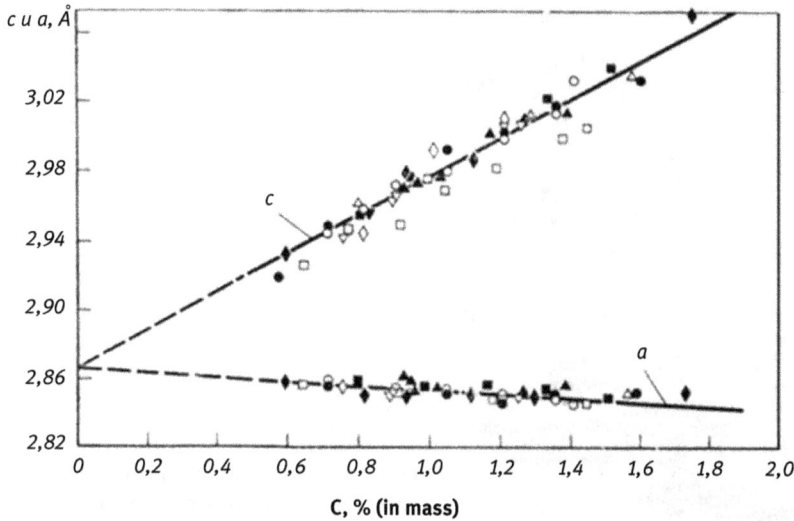

Figure 7.13: Dependence of martensitic lattice periods on the carbon content in steel. Numerous points are experimental data.

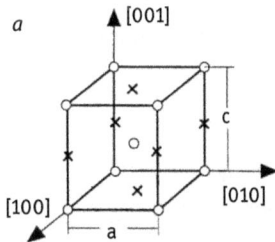

Figure 7.14: Arrangement of carbon atoms in the martensitic lattice. Crosses stand for possible positions of carbon atoms.

the crystals usually corresponds to the characteristic plane of habit. Let us examine four types of martensite: a lath (packet) type, a butterfly-like type, a lamellar (lenticular) type, and a fine-lamellar (low-temperature) type.

Lath martensite belongs to a large group of low- and medium-carbon steels whose martensitic point is located quite high (100–200 °C). Lath martensite crystals are in the shape of 0.2–2 μm thick laths. The size ratio of the laths in three dimensions is approximately 1:7:30. One-directed laths can be composed together and form packets (Figure 7.15a). There may be several packets within the original austenite grain. The laths contain a high density of dislocations (up to 10^{11} cm^{-2}) (Figure 7.15b).

Therefore, lath martensite is also called dislocation one. V.M. Schastlivtsev has ascertained that any set of the laths in the packet obeys certain laws. Firstly, a sufficiently large packet contains martensite laths of 6 out of 24 possible orientations. These orientations were obtained by a shear (the first Kurdjumov shear) along one of four possible planes of the {111} austenite type. This means that four types of

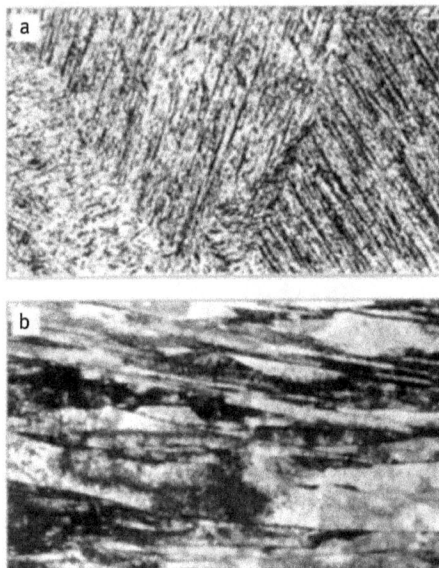

Figure 7.15: Structure of lath martensite in alloyed steel: (a) packets in the initial austenite grain ($\times 800$) and (b) laths in the packet ($\times 20,000$).

packets are possible to form. Secondly, the six orientations in the packet create three twin-bound pairs. Thirdly, the c-tetragonal axes of each pair of orientations include mutual angles of about 90°, which ensures a quasi-isotropic change in volume. These patterns are the result of minimizing the elastic energy that accumulates during martensitic transformation. Habitus of lath martensite is $\{111\}_A$ or close to it is $\{557\}_A$.

At lower martensitic point for alloyed steels with a content of 0.5–1.0% C, butterfly-like martensite is formed. In the early stages of transformation, aggregation of two separate lamellae or obtuse-angled butterfly-shaped pairs arises (Figure 7.16a). The transformation is completed by the formation of laths, and the butterflies become difficult to distinguish (Figure 7.16b).

The final microstructure consists of a mixture of lamellae (lamellar martensite) and laths. Lamellar martensite habitus is $\{225\}_A$. In high-alloyed alloys and high-carbon steels with an even lower martensitic point, lenticular (also lamellar) martensite is formed. Such martensitic crystals have the shape of biconvex lenses (Figure 7.17a) with a narrow, long region of thin twins (midribs) in their middle part (Figure 7.17b).

The peripheral parts of the crystals contain an increased dislocation density (up to 10^{10} cm^{-2}). The crystals are composed of truss- or lightning-shaped groups. The martensitic structure retains a significant amount of retained austenite. Habitus of lenticular martensite crystals is $\{259\}_A$.

In high-alloyed alloys, thin-lamellar martensite is formed at a very low martensitic point (below −120 °C) (Figure 7.18).

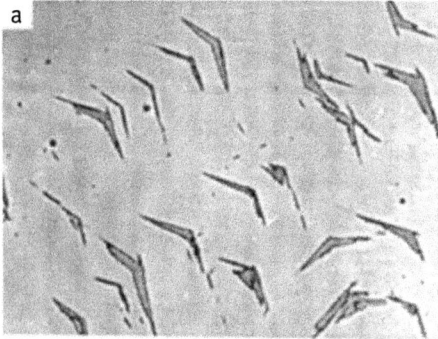

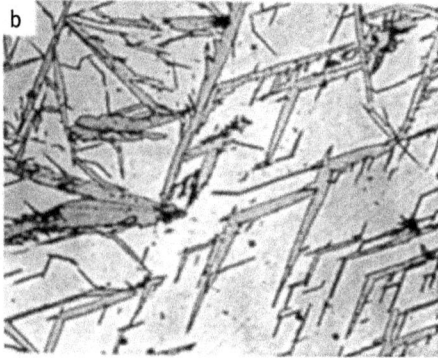

Figure 7.16: Structure of butterfly-like martensite($\times 150$): (a) start of transformation and (b) development of transformation.

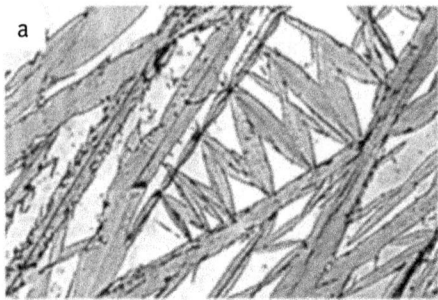

Figure 7.17: Structure of lenticular martensite: (a) microstructure ($\times 200$) and (b) region of a midrib ($\times 60,000$).

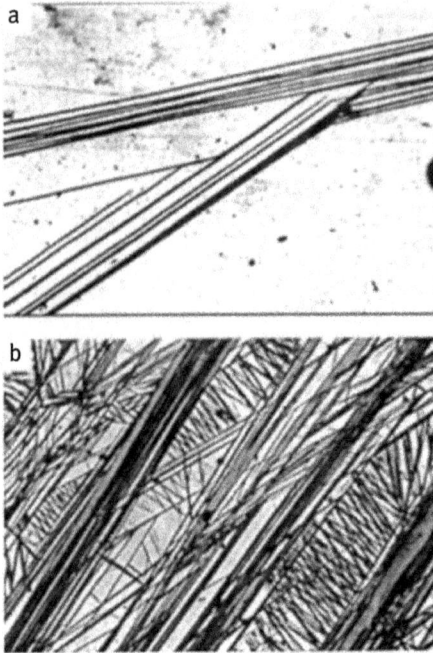

Figure 7.18: Thin-lamellar martensite: (a) start of transformation and (b) development of transformation (×200).

Thin, extended crystals often arise as clusters. Martensite crystals are fully twinned. After the transformation, a large amount of retained austenite is left. Habitus of thin-lamellar martensite is {3 10 15}$_A$.

As illustrated in Figure 7.19, a diagram of the temperature–concentration regions displays the existence of various types of martensite.

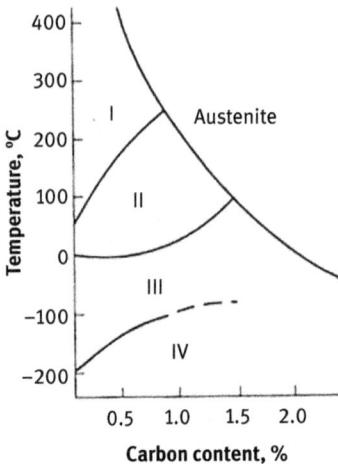

Figure 7.19: Dependence of the type of martensite on transformation temperature and carbon content in Fe–Ni–C alloys. Regions of martensite existence: I, lath martensite; II, butterfly-like martensite; III, lenticular martensite; IV, thin-lamellar martensite.

Figure 7.20 presents a stereographic triangle with schematic images of martensite types and the corresponding habit planes.

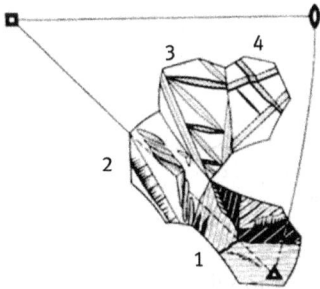

Figure 7.20: A stereographic triangle with characteristic regions of the arrangement of habit plane poles and schemes of the appropriate martensitic structures: 1, lath martensite; 2, butterfly-like martensite; 3, lenticular martensite; 4, thin-lamellar martensite.

Structural forms of martensite crystals are very diverse. This is explained by both a high level and ways of relaxation of the elastic stresses accompanying the martensitic transformation in steels and iron alloys. The relaxation mechanism acts as gliding and twinning at high and low transformation temperatures, respectively. Stress relaxation occurs largely due to certain ways of grouping of the crystals (packets and trusses).

Let us take a brief look at the influence of external impacts on the martensitic transformation. Since the latter is a deformation process, elastic stresses and plastic deformation are important factors affecting the transformation. The elastic stresses applied during cooling interact with the shape deformation to stimulate the nucleation and growth of martensitic crystals above the start temperature M_s. In this case, martensitic orientations are selected from all possible crystallographically equivalent ones. Those crystals are formed, whose shape strain is most favorably oriented relative to the direction of the applied stresses. This martensite is called stress-induced martensite. The martensitic transformation above the temperature M_s can also be caused by plastic deformation. This martensite is called strain-induced martensite. There is a certain point M_d close to the temperature T_0 (see Figure 7.9), above which plastic deformation no longer leads to martensitic transformation. Figure 7.21 shows a diagram of the influence of elastic and plastic deformation on the martensitic transformation temperatures.

Stress-induced martensite is formed near the start temperature M_s. At higher temperatures and at stresses above the yield strength of austenite, strain-induced martensite is formed. Stress-induced martensite crystals arise in the same areas of austenite where they would appear upon cooling without stress. Strain-induced martensite arises in new areas in which the deformation prepared nucleation sites.

The effect of the formation of martensite upon deformation has found practical application. TRIP (transformation-induced plasticity) steels were created, whose plasticity characteristics can be enhanced under a certain treatment. This is possible to do due to the occurrence of martensitic transformation during deformation. Thus, the martensitic

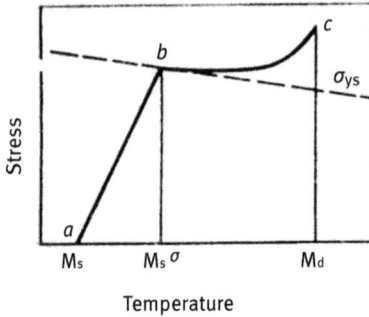

Figure 7.21: Martensite of stresses (*ab* segment) and martensite of deformations (*bc* segment). σ_{ys}, yield strength of austenite.

transformation came as an additional deformation channel to gliding and twinning. There is another feature of TRIP steels to form martensite during plastic deformation. It consists in creating local hardening, and the plastic flow is transferred to unhardened areas. This effect prevents from nucleating a crack (and its growth, if it has arisen) and increases the total deformation. The issue on TRIP steels is encountered in the specialized literature.

At a temperature above M_d, preliminary plastic deformation of austenite followed by cooling affects the martensitic transformation. Small-degree deformation activates the transformation during cooling. Large-degree deformation decelerates the martensitic transformation.

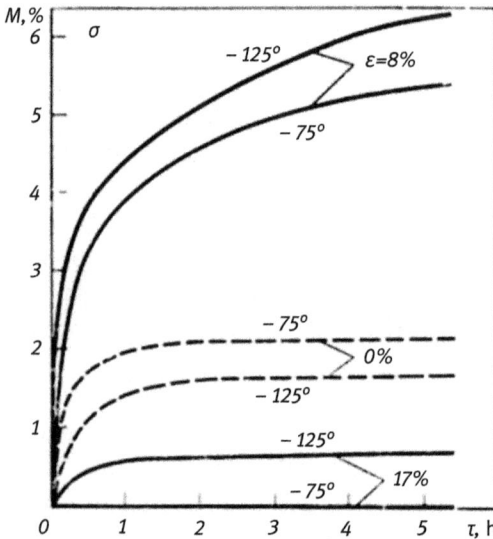

Figure 7.22: Activating and inhibiting influence of preliminary plastic deformation at +100 °C on martensitic transformation during isothermal holdings at temperatures of −75 and −125 °C.

Figure 7.22 shows the change in the amount of martensite during isothermal holding (for two temperatures) in Cr17Ni9 steel after 8%- and 17%-degree deformation of
austenite, respectively. Deformation by 8% increases the rate of isothermal transformation and the total amount of martensite formed. During deformation by 17%, both
the rate of the formation of martensite and its amount decline. Small deformation is
thought to prepare sites for martensitic crystals to nucleate. Large deformations im-
pede the growth of the crystals.

Scientists at the Institute of Metal Physics have discovered the initiating influence of a magnetic field on the martensitic transformation. V.D. Sadovsky, a full
member of RAS, and his colleagues have found that a high-intensity pulsed magnetic
field (100–300 kOe) makes non-ferromagnetic austenite to convert into ferromagnetic
martensite.

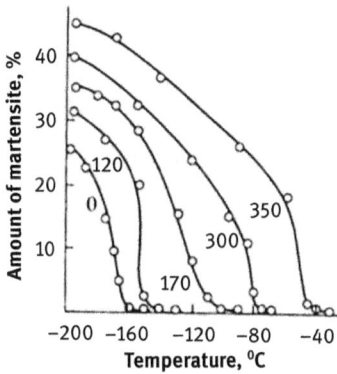

Figure 7.23: Influence of the strength of a pulsed
magnetic field on the martensite range position in high-
alloyed Cr–Ni steel. The numbers near the curves are
indicated in kOe.

Figure 7.23 illustrates how the temperature range of the martensitic transformation
in 50Cr2Ni22 steel is shifted under the influence of a pulsed magnetic field of various strengths. Exposed to an external field of 350 kOe, the martensitic point goes
up by more than 100°, and the amount of martensite almost doubles. The action of
the magnetic field is well described by the Krivoglaz–Sadovsky formula:

$$\Delta T = T_0(V_1 M_1 - V_2 M_2)Hq^{-1},$$

where ΔT is the shift of the phase equilibrium temperature (martensite and austenite)
under the influence of a magnetic field, T_0 is the phase equilibrium temperature without the action of a field, V_1 and V_2 are the molar volumes of the phases, M_1 and M_2 are
the magnetic moments of the phases, H is the magnetic field strength, and q is thermal
effect of transformation. In the case when one of the phases is non-ferromagnetic,
such as austenite in low-alloyed steels, that is, $M_2 = 0$, then the formula is simplified:

$$\Delta T = T_0 VMHq^{-1},$$

where V is the molar volume, and M is the magnetic momentum of the magnetic phase. Studying the influence of a magnetic field on the martensitic transformation has made it possible to elucidate some specific transformation features such as the relationship between athermal and isothermal transformations and the growth stages of lenticular martensite crystals.

At the beginning of this section, it was said that the position of the martensitic point is cooling rate independent over a wide range of rates. D.A. Mirzaev and O.P. Morozov have revealed that the position of the martensitic point is stepwise dependent on specially created, extremely large cooling rates (up to 10^6 K/s).

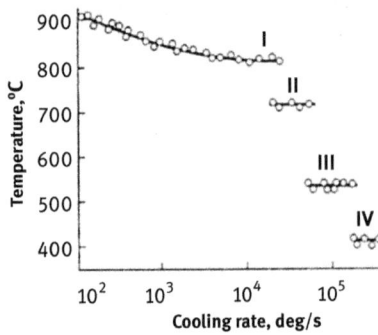

Figure 7.24: Influence of the cooling rate on temperature of the $\gamma \rightarrow \alpha$ transformation in iron with 0.005% C. I–IV, transformation stages.

As shown in Figure 7.24, there are four stages of austenite transformation found in iron with 0.005% C at high cooling rates, three of which are martensitic ones. The cause of the stepping effect is believed to be associated with various mechanisms of stress relaxation. Studies of such dependencies for steels with various alloying systems have discovered new regularities of austenite transformations in steels.

7.6 Intermediate (bainitic) transformation

Bainite is a structure formed as a result of bainitic transformation. In general, bainite consists of three phases: bainitic ferrite, cementite, and retained austenite. In silicon-alloyed steels, the formation of cementite is suppressed, and bainite consists of ferrite and austenite (carbide-free bainite).

Bainitic transformation occurs in an intermediate temperature range, between the temperatures of pearlite and martensitic transformations. As to bainitic and pearlitic transformations, C-shaped decomposition curves for supercooled austenite can share the same curve (see Figure 7.2) and be split into two branches (see Figure 7.3). In carbon steels, the temperature range of bainitic transformation is approximately 500–250 °C. Upper and lower bainite are distinguished by the formation temperatures and the type of structure obtained (see below). The transition between the upper and lower bainite lies at about 350 °C.

Bainitic transformation is intermediate between pearlite and martensitic ones not only in terms of transformation temperatures but also in its characteristics. The comparative parameters of pearlitic, bainitic, and martensitic transformations are given in Table 7.1.

Table 7.1: Characteristics of pearlitic, bainitic, and martensitic transformations.

Parameter	Pearlite (P)	Bainite (B)	Martensite (M)
Free energy of the conversion product, F	F_P	$F_B > F_P$	$F_M > F_P$
Formation temperature, T (°C)	$T_A > T_P > {\sim}500\ °C$	${\sim}500\ °C > T_B > M_S$	$M_S \geq T_M$
Reaction front rate, G_{max}	G_P	$G_B \approx G_P$	$G_M >> G_P$
Deformation of the shape, pattern on the polished surface	No	Yes	Yes
Carbon diffusion paths, S (nm)	$S_P > 10$	$10 > S_B > 0$	$S_M = 0$
Strength of the conversion product, σ	σ_P	$\sigma_B > \sigma_P$	$\sigma_M > \sigma_P$

The formation of bainite, as well as martensite, is accompanied by the appearance of a characteristic pattern on the polished surface of thin sections (Figure 7.25).

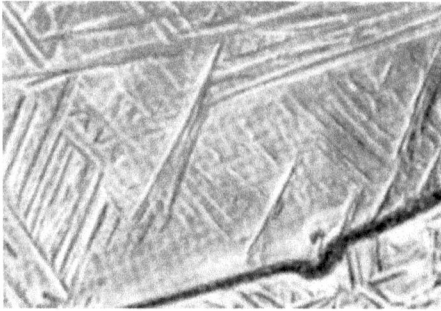

Figure 7.25: Pattern on the polished surface of a thin section during bainitic transformation at 490 °C in steel with 0.9%C. The image is obtained at 490 °C.

The pattern indicates the shape deformation and, as a consequence, the shear mechanism of transformation. However, the maximum growth rate of bainitic ferrite crystals is much lower than that of martensite and is close to the growth rate of a pearlite colony. This is the dual ("intermediate") nature of the bainitic transformation.

As is clear from the table, the values of the carbon diffusion path of the bainitic transformation lie between those of the pearlitic and martensitic transformations.

Bainitic transformation occurs at temperatures below 500 °C, when the diffusion rate of substitution atoms (e.g., iron) is low, and the diffusion rate of interstitial atoms (carbon and nitrogen) is quite high. These temperatures exclude the possibility of normal diffusion formation of ferrite. However, the latter can be formed by a shear mechanism at temperatures below T_0 (see Figure 7.10). In order for the bainitic transformation to begin in steels of a given chemical composition above the temperature M_s, local fluctuation carbon depletion is necessary. In the depleted region, the temperature M_s rises, and a bainitic ferrite nucleus (shear mechanism) arises. For the nucleus to grow, the removal of carbon atoms from the interphase boundary into austenite is required. Therefore, the growth rate is limited by the carbon diffusion rate in austenite. Measurements have shown that the activation energies of crystal growth of bainitic ferrite and carbon diffusion in austenite are of the same order of magnitude (12,000–17,000 and 32,000–36,000 cal/g-atom, respectively). Thus, the mechanism of bainitic transformation includes both shear rearrangement of the austenite lattice into a ferrite lattice and diffusion removal of carbon from the boundary of the growing ferrite crystal. The diffusion redistribution of carbon initiates the appearance of weakly carbon-supersaturated ferrite and carbon-enriched austenite. The formation of cementite and the preservation of retained austenite in the bainitic structure are secondary processes (R.I. Entin).

Bainitic ferrite crystals are usually lath-shaped, similar to martensite ones. Laths of upper bainite are larger than those of lower ones. The lath dislocation density is high but lesser than that in lath martensite. The carbon concentration in ferrite of the lower bainite is greater than in the upper ferrite. The precipitation of cementite particles in the lower bainite occurs inside the laths. In the upper bainite, cementite is released along the boundaries of the laths (Figure 7.26).

This arrangement of cementite particles points to the precipitation of cementite inside ferrite or austenite during the formation of either the lower or upper bainite, respectively.

Due to carbon enrichment in the process of bainitic transformation, austenite has increased resistance to martensitic transformation. When cooled from the transformation temperature to room temperature, austenite can partially turn into martensite or remain unconverted. The transformation temperature is responsible for the completeness of the bainitic transformation (and the amount of unconverted austenite). Figure 7.27 shows how the degree of transformation varies for different transformation temperatures.

For comparison, a similar dependence is given for the martensitic transformation. In bainitic transformation, the amount of retained austenite is large, and it can reach 40% in alloyed steels. In upper bainite, the retained austenite has the appearance of irregularly shaped sections. In lower bainite, it is located along the boundaries of ferrite laths (Figure 7.28), as in lath martensite.

The carbon content in austenite is determined by two opposite processes. The former is the redistribution of carbon between ferrite and austenite. In this case, the

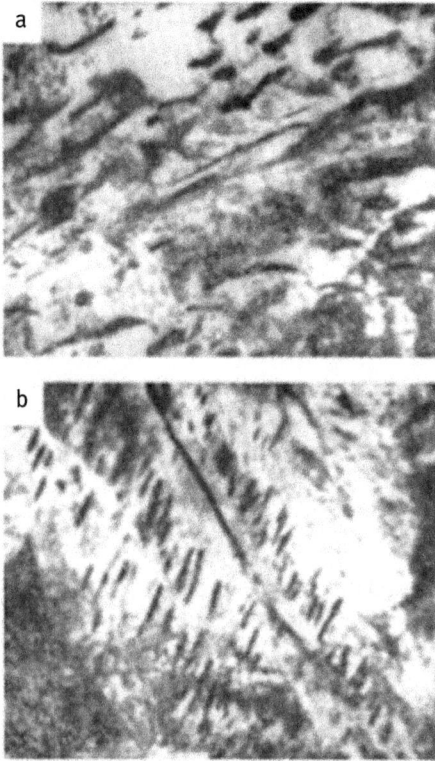

Figure 7.26: Carbides in upper (a) and lower (b) bainite ($\times 24,000$).

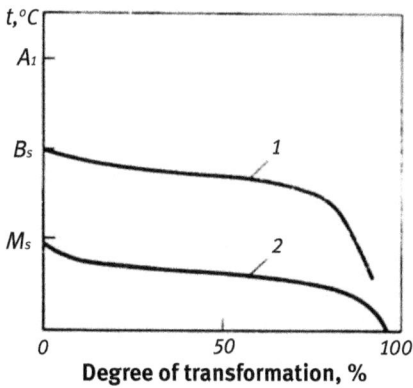

Figure 7.27: Temperature influence on the transformation degree of austenite during bainitic (1) and martensitic (2) transformations.

carbon content in austenite increases. The latter is the precipitation of carbides, in which the carbon content decreases. Figure 7.29 presents three cases of changes in carbon concentration during isothermal holding (R.I. Entin).

The course of dependencies is determined by the predominance of a particular process. Curve 1 corresponds to the diffusion carbon enrichment process in austenite.

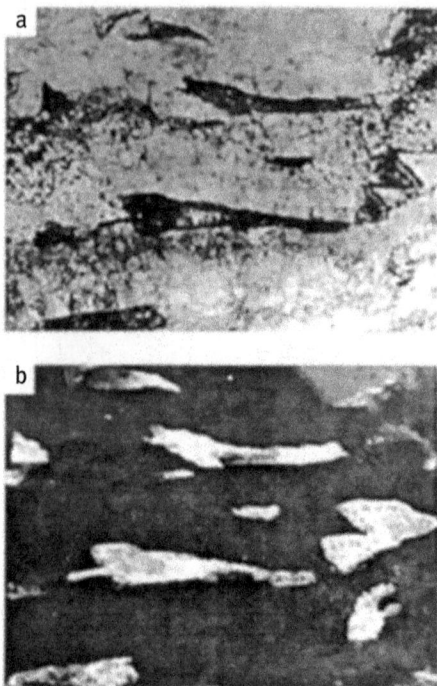

Figure 7.28: Retained austenite in lower bainite($\times 30,000$): (a) structure of retained austenite and (b) dark-field image in the reflection of austenite.

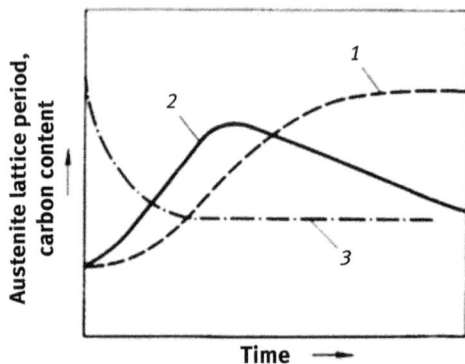

Figure 7.29: Scheme of change in carbon concentration in austenite during isothermal holding.

Curve 3 corresponds to the carbon depletion process due to the precipitation of carbides. Curve 2 reflects the carbon enrichment process in austenite at the beginning of the bainitic transformation, followed by the carbide precipitation to isolate carbon from the solid solution. Which of these processes will be implemented depends on a number of factors such as the chemical composition of the steel (primarily on the carbon content), and the temperature and duration of the transformation.

Bainite, like martensite, has an orientational relationship with the original austenite. Mutual orientations of bainitic ferrite and austenite correspond to the Nishiyama orientation relationships:

$$(011)_F \parallel (111)_A \text{ and } [0\bar{1}1]_F \parallel [11\bar{2}]_A$$

or intermediate relationships between the Nishiyama and Kurdjumov–Sachs orientation relationships. In other words, they do not differ from the relationships during
martensitic transformations. With the shear mechanism of bainitic transformation,
such an orientation relationship is expected. In upper bainite, the orientation relationship between cementite and ferrite can be either the Petch relationship or the
Bagaryatsky relationship, as in pearlite. In lower bainite, the orientation relationship between cementite and ferrite

$$[100]_C \parallel <0\bar{1}1>_F, \quad [010]_C \parallel <1\bar{1}\bar{1}>_F, \quad [001]_C \parallel <211>_F,$$

either coincides with the Bagaryatsky orientation relationship or slightly differs
from the latter approaching the Isaichev orientation relationship.

It should be noted that D.A. Mirzaev and V.M. Schastlivtsev have put forward a
new concept of bainitic transformation, which connects the formation of bainite and
stepwise martensitic transformation at ultrahigh cooling rates. This concept has not
yet found widespread use but it can be found in the specialized literature.

Bainitic transformation can occur not only in steel but also in non-ferrous alloys
that are conditional upon a great difference in diffusion mobility of atoms of the components. A typical example is a Cu–Zn (brass) system, in which the diffusion rate of
zinc is much greater than that of copper. In brass Cu–41% Zn, a body-centered β-
phase transforms into a face-centered α-phase at 350 °C. The transformation is caused
by a shear mechanism (with the appearance of a characteristic pattern) due to the
slow growth of thin α-phase lamellae. The initial β-phase is enriched with zinc that
plays a role similar to that of carbon in steels. The orientational coupling of the lattices
of the α- and β-phases is identical to martensitic transformation.

7.7 Tempering of hardened steel

Against the above-described transformations of supercooled austenite, tempering
of hardened steel is not a transformation but a heat treatment operation, including
several transformations. Their proceeding during tempering depends on the heating
temperature and the holding time. There is low (temperatures of 120–250 °C), medium
(350–450 °C), and high (450–650 °C) tempering.

Processes that occur during tempering of hardened carbon steel bring the initial
metastable state closer to equilibrium. The latter stands for the coexistence of cementite and ferrite with a carbon concentration of less than 0.02 wt% (see Figure 7.1). In

alloyed steels, the slow formation of special carbides prevents from approaching the equilibrium. The structure of the hardened steel consists of a carbon-supersaturated solid solution (martensite) with a body-centered tetragonal lattice and some amount of retained austenite. The microstructure of martensite is characterized by a large number of boundaries/sub-boundaries between crystal laths and a high dislocation density. These structural features affect tempering. As the heating temperature rises, transformations in hardened steel pass in the following sequence:

1. The redistribution of carbon atoms in a solid solution. The carbon atoms move to dislocations and boundaries, redistributing among the interstices, and possibly to form carbon clusters.
2. The decomposition of martensite to form first carbon-enriched areas, then carbides precipitate. The carbides' structure and composition and their relationship with the matrix change.
3. The decomposition of retained austenite to form pearlite or bainite. Austenite turns into martensite upon cooling below the tempering temperature.
4. Precipitation of dispersed particles of special carbides. The structure and composition of carbides in alloyed steels change.
5. Relaxation of internal stresses over the entire tempering temperature range. The fine structure of martensite changes till recrystallization at high tempering happens.
6. Coagulation of carbides, and redistribution of alloying elements and impurities throughout the volume and grain boundaries.

It is worth emphasizing that the temperature ranges of transformations overlap. The main transformation during tempering is the decomposition of martensite to form carbides. Dilatometric (Hannemann and Trager) and X-ray (Kurdjumov) investigations of hardened high-carbon steel at heating have shown that tempering can be divided into three stages. At temperatures of 100–150 °C, the samples become shorter, and the periods of the tetragonal lattice of martensite vary. This fact indicates the removal of carbon from the solid solution. At 150–300 °C, the samples elongate, and at 260 °C the X-ray reflections of austenite disappear. This means that the retained austenite decomposes. At 300–350 °C, the length of the samples again decreases, and X-ray reflections of cementite appear; martensite continues decomposing till precipitation of carbide of the cementite type comes. Heat capacity curves of hardened steels with different carbon contents (Figure 7.30) also point to the staging of the transformations during tempering.

However, in steels with a carbon content of less than 0.41% C, the first stage is absent. Studies of heat capacity have shown (see Figure 7.30) that ~ 0.2% C remains in martensite after the first-stage tempering (up to a heating temperature of 250 °C). Therefore, with an increase in the tempering temperature to 300 °C or higher, the decomposition of martensite continues.

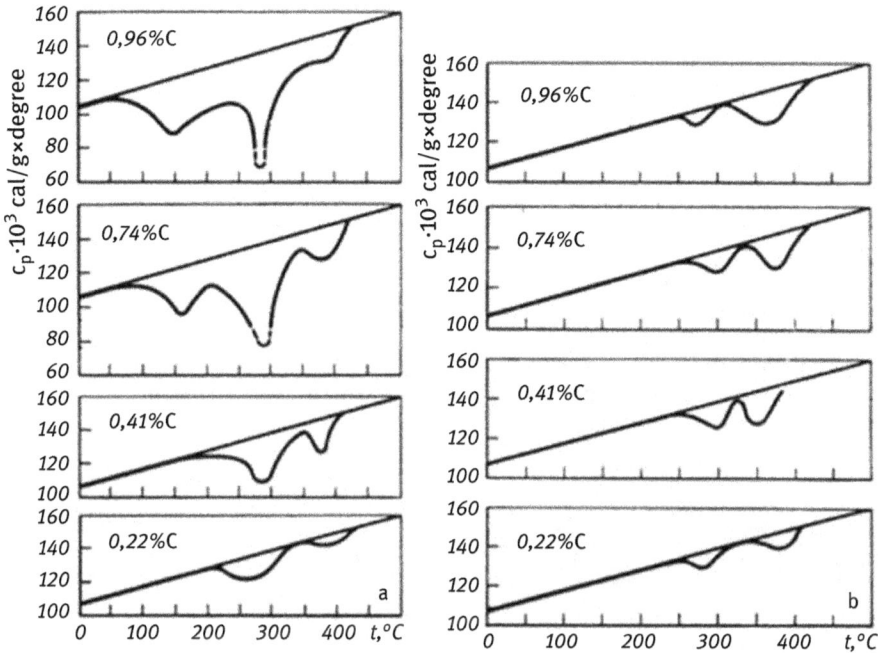

Figure 7.30: Heat capacity of quenched steels with a different carbon content: left patterns, at heating; right patterns, after preliminary tempering at 250 °C.

The precipitation of cementitious carbides takes place in several stages, dictated by the rule of steps. At low temperatures (up to 150 °C), ε-carbide with a hexagonal lattice arises, designated as ~ Fe_2C. The Curie temperature of this carbide (~ 380 °C) is significantly higher than that of cementite (210 °C). A decrease in the surface energy at the carbide/martensite boundaries (compared with the cementite/martensite ones) contributes to the formation of metastable ε-carbide. This fact can explain the better connection of the ε-carbide crystal lattice with the martensite lattice, unlike the cementite lattice. There is evidence that metastable carbide as χ-carbide with a monoclinic lattice may form in high-carbon steels. Its composition corresponds to ~ Fe_5C_2, and the Curie temperature is ~ 270 °C.

At temperatures of 200–300 °C, rhombic-lattice cementite Fe_3C appears in the steel structure according to the state diagram (see Figure 7.1).

At 350–400 °C, its formation finishes. Cementite precipitates inside martensite laths in the shape of thin plates of two or three orientations (Figure 7.31a). In lamellar martensite crystals, cementite precipitates at the boundaries of laths and twins in the shape of plates of the same orientation (Figure 7.31b). With high tempering, carbide precipitates are transformed into globular particles (Figure 7.31c). It is established that cementite can be precipitated directly from a supersaturated solid

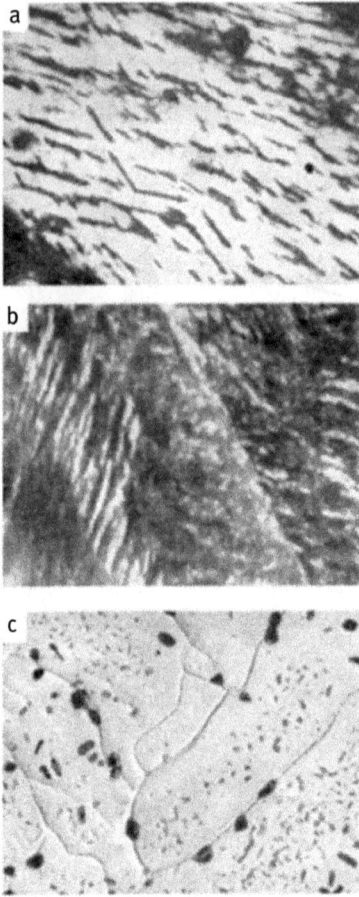

Figure 7.31: Three types of carbides in medium-carbon alloyed steels: (a) cementite of two orientations inside laths, tempering at 325 °C ($\times 30,000$); (b) cementite at the boundaries between laths and twins. Tempering at 300 °C. A dark-field image($\times 45,000$): (c) carbides precipitate at the boundaries and inside grains. Tempering at 650 °C + 500 °C, 300 h, carbon replica ($\times 8,000$).

solution or by rearrangement of the metastable carbide lattice into a cementite lattice. In the first case, particles of metastable carbide previously formed get dissolved.

As in pearlite transformation, there is an orientational relationship between the lattices of cementite, martensite, and austenite. It is expressed either by the Isaichev orientation relationships:

$$(013)_C \| (011)_M \| (111)_A,$$

$$[010]_C \| [1\bar{1}1]_M \| [1\bar{1}0]_A.$$

or the orientation relationships of Bagaryatsky:

$$[100]_C \| [110]_M, \quad [010]_C \| [1\bar{1}1]_M, \quad [001]_C \| [\bar{1}12]_M.$$

The lattice of metastable ε-carbide is connected with the martensitic lattice by the relationships known as Jack relationships:

$$(0001)_\varepsilon \parallel (011)_M \text{ and } [100]_\varepsilon \parallel [1\bar{1}1]_M.$$

More detailed information on the orientational coupling of carbide and martensite lattices can be found in the specialized literature.

At high tempering temperatures (450 °C and higher), the diffusion mobility of substitution atoms (iron and alloying elements) increases, and special carbides form in alloyed steels. They emerge either due to the gradual enrichment of cementite particles with alloying elements followed by rearrangement of the crystal lattice or due to direct nucleation at dislocations and boundaries/sub-boundaries. In both cases, the formation of special carbides obeys the rule of steps. So, in chrome steels, the formation of carbide phases occurs in the following sequence:

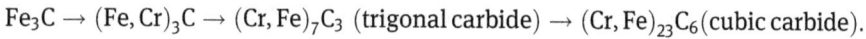

$$Fe_3C \rightarrow (Fe, Cr)_3C \rightarrow (Cr, Fe)_7C_3 \text{ (trigonal carbide)} \rightarrow (Cr, Fe)_{23}C_6 \text{(cubic carbide)}.$$

For steels alloyed with molybdenum and tungsten, the sequence of carbide formation is as follows:

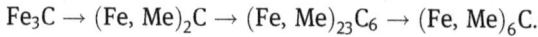

$$Fe_3C \rightarrow (Fe, Me)_2C \rightarrow (Fe, Me)_{23}C_6 \rightarrow (Fe, Me)_6C.$$

It is seen that the sequential rearrangement of the carbide phases increases the content of the alloying (carbide-forming) element, in accordance with the rule of steps. It should be underscored that special-carbide particles are much smaller than cementite particles.

> The main reason is the low diffusion mobility of atoms of alloying elements. Thus, unlike completing carbide formation in carbon steel at medium and high tempering and carbon removal from a supersaturated solid solution, followed by only coagulation of cementite, the decomposition of martensite in alloyed steels includes only the initial, relatively short-term stage of carbide formation, which is followed by a prolonged diffusion redistribution of alloying elements between ferrite and the carbide phase, the enrichment of the latter with carbide-forming elements, or even the conversion of some carbides to others. (Kurdjumov, Entin)

During prolonged tempering at high temperatures, along with the coagulation of cementite, the ferrite matrix recrystallizes (Figure 7.31c). However, the lath structure obtained during the martensitic transformation is very stable (Figure 7.32).

Dispersed particles of special carbides in alloyed steels inhibit recrystallization, and the lath structure preserves up to high tempering temperatures, close to the temperature of austenite formation. The most effective carbides are vanadium, niobium, and titanium.

As mentioned above, retained austenite decomposes when tempered. The decomposition kinetics is described by C-shaped curves, similar to those for supercooled austenite. In this case, the incubation period of the decomposition of retained austenite is lesser, and the decomposition rate is greater than that for supercooled austenite. Depending on the tempering temperature and chemical composition of steel, the retained austenite decomposes on either pearlite or bainite. At high tempering temperatures,

Figure 7.32: Lath structure and coalescence of cementite at high tempering ($\times 27,000$).

austenite is carbon-depleted due to the precipitation of carbides. The martensitic point goes upward, and a martensitic transformation occurs in the retained austenite upon cooling below the tempering temperature. It holds if the retained austenite has not completely transformed at high temperatures.

Quenched steel has high hardness and strength but low ductility and fracture toughness (toughness). Therefore, steel after quenching must be subjected to low or high tempering, depending on practical purposes. Quenched steel when tempered declines its strength characteristics, but enhances its ductility and toughness. Figure 7.33 outlines the change in the hardness of quenched carbon steels as a function of tempering temperature.

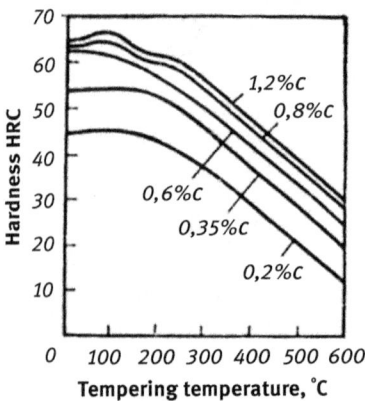

Figure 7.33: Tempering–temperature dependence of hardness of carbon steels of a different composition.

The drop in hardness during tempering is due to the decomposition of martensite as a solid and brittle structural component and the formation of low-carbon ferrite.

Tempering in certain temperature ranges leads to a sharp decrease in toughness. This phenomenon is called temper embrittlement. Let us briefly dwell on the description of this fact and structural reasons for its emergence. Figure 7.34 sketches the tempering–temperature dependence of the impact toughness measured at 20 °C for quenched 37ChNi3A chromium–nickel steel.

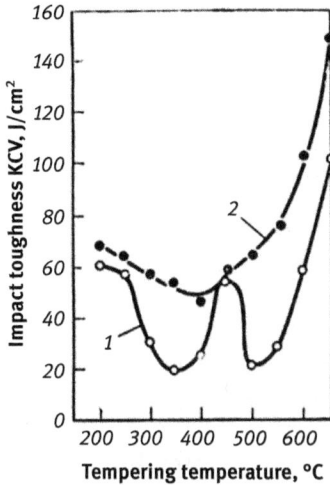

Figure 7.34: Tempering–temperature dependence of impact toughness of medium-carbon Cr–Ni steel: 1, after quenching and tempering upon rapid cooling; 2, after high-temperature thermomechanical treatment.

Curve 1 has two "dips" in the temperature ranges of 250–400 °C and 450–600 °C. These temperatures match irreversible and reversible temper embrittlement, respectively. The former has a minimum value of toughness at 350 °C. At this temperature, the decomposition of retained austenite is usually completed, and carbides precipitate in the form of interlayers at the boundaries of either grains or laths. The most probable cause of irreversible temper embrittlement is the destruction of a long structural part weakened by the carbide phase. Carbides can serve as centers to initiate brittle cracks and contribute to their development. The destruction can be both inter- and intra-crystalline. The latter case is typical of high-purity steels. As to commercial steels, the segregation of various impurity elements (primarily phosphorus) at grain boundaries may aggravate the development of irreversible temper embrittlement. The nature of this type of brittleness is not fully understood yet, and remedies of its elimination have not been found. Therefore, tempering at temperatures of 300–400 °C (medium-temperature tempering), as a rule, is not used.

Reversible temper embrittlement can be detected during slow cooling in a dangerous temperature range (550–500 °C) or during tempering in this range. After being heated above the dangerous temperature range and cooled quickly, steel in a state of reversible temper embrittlement loses its brittleness. However, the steel re-heated at the dangerous temperature range becomes brittle again. Therefore, this type of brittleness is called reversible.

The nature of the destruction of steel being in a state of reversible temper embrittlement is brittle, inter-crystalline. A measure of reversible temper embrittlement is an increase in the temperature of the viscous–brittle transition (cold brittleness threshold) upon impact toughness tests. The brittleness threshold is determined by the test–temperature–impact toughness plot. Such a dependence is called a serial curve.

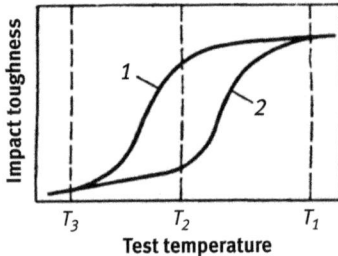

Figure 7.35: Test–temperature dependence of impact toughness: 1, temper embrittlement is absent; 2, temper embrittlement state.

Figure 7.35 exemplifies the arrangement of two serial curves of steel in a state of reversible temper embrittlement (curve 2) and without it (curve 1). The impact toughness at test temperatures T_1 and T_3 is the same for the two states. Therefore, reversible temper embrittlement is impossible to reveal in these cases. However, the impact toughness of steel in the state of reversible temper embrittlement decreases sharply at T_2. Such temperature corresponds to measurements at 20 °C, shown in Figure 7.34.

The nature of reversible temper embrittlement is largely elucidated. Auger spectroscopy is used to find segregations of such impurities as phosphorus, antimony, tin, and arsenic at the boundaries of the initial austenitic grains. These impurities are redistributed between the volume and grain boundaries during tempering within a dangerous temperature range. The kinetics of segregation over temperature–time regimes is established to correspond to that of temper embrittlement observed. These experiments confirm Arkharov's concept of inter-crystalline internal adsorption. According to V.I. Arkharov, some impurities enrich grain boundaries (horophilic impurities), while others go into the grain volume (horophobic). The segregations of phosphorus, antimony, tin, and arsenic (horophilic) weaken the inter-granular adhesion, and a fragile, inter-granular nature substitutes for the viscous destruction.

For a visual representation of the phenomenon of reversible temper embrittlement, the Ioffe scheme should be employed (Figure 7.36).

The yield strength (S_{ys}) increases dramatically with decreasing temperature; the resistance to brittle separation (S_O) increases slightly. The intersection of the lines S_{ys} and S_O (curves 1 and 2) indicates the temperature of the viscous–brittle transition T'_x. Below this temperature, the viscous destruction is replaced by the brittle one. In a state of reversible temper embrittlement, the resistance to brittle inter-granular destruction drops (curve 3). In this case, the corresponding lines S_{ys} and S_O intersect at a higher temperature $T''_x (T''_x > T'_x)$, that is, the temperature of the viscous–brittle transition rises. With constant grain boundary strength, an increase in the yield strength (curve 4) enhances the temper embrittlement $(T'''_x > T''_x)$.

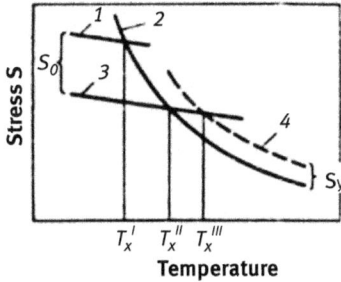

Figure 7.36: Temperature curves of yield strength S_{ys} and resistance to brittle destruction S_0 (the Ioffe scheme).

Knowledge of the nature of reversible temper embrittlement allows one to find ways to suppress this phenomenon. Firstly, it is necessary to reduce the content of detrimental impurities. This primarily relates to phosphorus. The concentration of the latter in industrial steels amounts to several hundredths of a percent. With segregation, its concentration at the grain boundaries can increase by two orders of magnitude. Secondly, the tempering temperature must be beyond a dangerous temperature range. With high-temperature tempering, cooling in the dangerous temperature range needs to be fast enough to prevent segregation. Thirdly, little additives of molybdenum (0.2–0.4%) or tungsten are known to delay phosphorus segregation, reducing the tendency of steel to reversible temper embrittlement. Fourthly, high-temperature thermomechanical treatment enlarges the total grain boundary area, thereby diminishing the content of detrimental impurities per unit area and weakening the temper embrittlement (see Figure 7.34).

7.8 Austenite formation

When heated, austenite to be created exhibits its kinetics through the rates of nucleation and growth. Its kinetics is similar to pearlite transformation during heating. The number of nuclei n arising in a unit volume per unit time is expressed as the product of two exponentials:

$$n = K \exp\left(-\Delta\Phi/kT\right) \exp\left(-U/kT\right),$$

where $\Delta\Phi$ is the work to form a critical sized nucleus, U is the activation energy of the transition of atoms to the nucleus, and K is a proportionality coefficient. Both exponentials increase with increasing temperature. The same dependence is typical of the growth rate (Figure 7.37).

Therefore, with rising temperature and, as a consequence, the degree of overheating above the phase equilibrium temperature, the rates of nucleation and growth of austenite continuously go up. Consequently, the total transformation rate increases always and reaches not maximum, unlike the pearlite transformation.

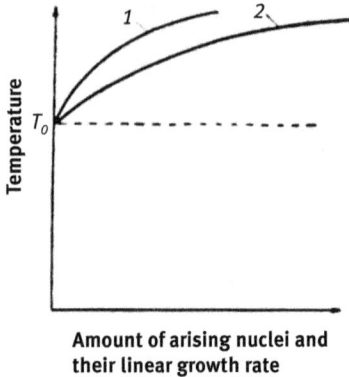

Amount of arising nuclei and their linear growth rate

Figure 7.37: Heating temperature dependencies of the growth rate (1) and nucleation number (2). T_0, equilibrium temperature.

Let us consider the formation of austenite as the initial structure of ferrite + cementite. According to the state diagram of iron–cementite (see Figure 7.1), at temperatures above the eutectoid line, austenite is formed from low-carbon ferrite with an increase in carbon concentration. Therefore, the most favorable austenite nucleation sites are ferrite/cementite interphase boundaries, boundaries of pearlite colonies, grain boundaries, and sub-boundaries in ferrite, where the carbon concentration is higher. Experimental data indicate that there are two stages to form austenitic sites. The former is the polymorphic conversion of ferrite to austenite with the formation of metastable low-carbon austenite. The latter is an increase in the concentration of carbon in austenite. Otherwise speaking, austenite approaches the equilibrium composition due to the dissolution of carbide particles in it. The equilibrium carbon concentration in austenite is determined by the *GS* line of the iron–cementite diagram (see Figure 7.1).

According to M.A. Stremel, the austenite nucleus arises by a shear mechanism, and its crystal lattice is coherently bound to the ferrite lattice. Coherent conjugation of the lattices ensures the minimization of the surface energy of the interphase boundary formed. In the process of nucleus growth, the coherent bond disappears, and the shear mechanism is replaced by the normal growth mechanism of austenitic grains.

The formation of austenite includes the following processes: the emergence of austenite nuclei and their growth due to ferrite and cementite, the disappearance of ferrite, the dissolution of cementite, and the homogenization of austenite. The highest growth rate of austenite sites is typical of the places where ferrite and cementite come into contact. This is because diffusion paths here are the smallest. In this case, two different initial structures of ferrite + cementite are necessary to consider the growth of austenite. In one case, globular particles of cementite are distributed throughout the ferrite; in the other case, the cementite of pearlite is lamella-shaped. The former involves the enveloping of the globular particles of cementite by a layer of austenite when they grow. As the austenitic shell thickens, carbon diffusion paths become longer, and the growth rate of austenitic sites diminishes. Bulk carbon diffusion controls the growth in austenite. The process of completion of austenitization can be delayed conditional upon a large size of cementite particles. The latter determines the carbon diffusion paths as an inter-lamellar distance in pearlite, which remains unchanged during the absorption of pearlite by austenite. Figure 7.38 shows the displacement of the front of a growing austenite site along a pearlite colony.

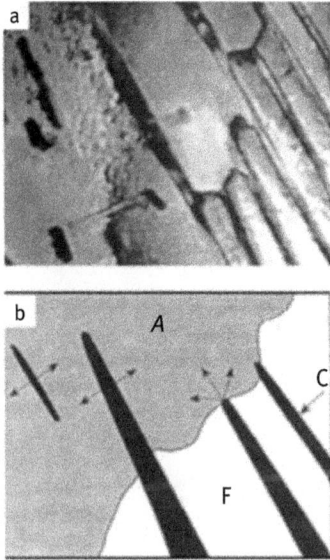

Figure 7.38: Growth of austenite in the pearlite structure: (a) structure ($\times 45,000$) and (b) scheme.

The arrows point to the carbon diffusion directions. The austenitic region contains incompletely dissolved carbide particles. Carbon diffusion is also responsible for the growth, but the diffusion occurs along a moving boundary. The diffusion paths are shorter than in the first case and do not increase during growth. Therefore, the austenitization completion process in the pearlite structure proceeds faster than in ferrite with granular cementite. The kinetics diagram of the isothermal formation of austenite in pre-eutectoid steel with a ferrite–pearlite structure (see Figure 7.4) shows that unconverted ferrite is left after the transformation of pearlite into austenite. After the conversion of ferrite, carbides are retained in the structure. It is worth emphasizing that after the dissolution of carbides, the austenite obtained is not uniform in chemical composition. Therefore, additional holding is required to homogenize it.

Figure 7.39 illustrates the peculiarities of the growth kinetics of two austenitic sites originating in pre-eutectoid steel at different boundaries (S.S. Dyachenko).

Region 1 arises near a pearlite colony. Region 2 emerges at the ferrite/ferrite boundary. The carbon concentration in region 1 is 0.53%; in region 2 it amounts to 0.18%. Region 1 is located closer to the carbon source (pearlite cementite) than region 2. Therefore, the former is more rapidly saturated with carbon.

Austenitization is completed by the formation of a new grain microstructure. New grains, as a rule, are not bound with the initial austenitic grains. This means

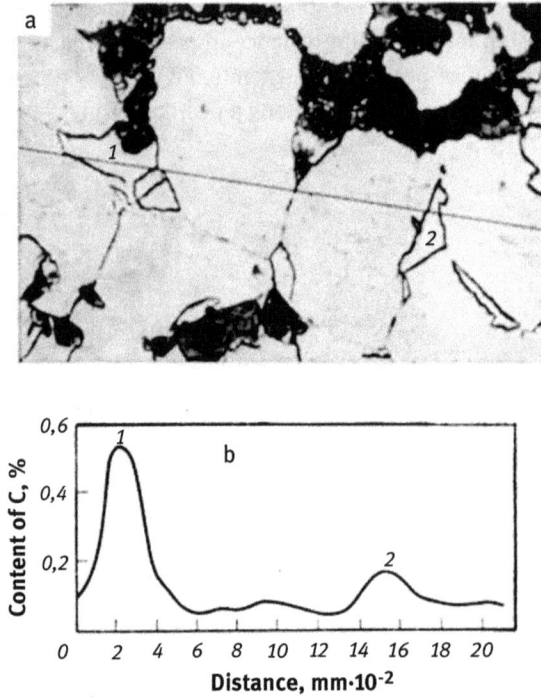

Figure 7.39: Austenite formation in pre-eutectoid steel: (a) microstructure and (b) carbon distribution along a line after a partial transformation.

that the phase transformation of the ferrite–cementite structure into austenite is accompanied by structural recrystallization (a change in the grain structure).

A different situation unfolds when steel with the initial martensitic (or bainitic) structure is transformed into austenite. In this case, structural heredity may manifest, and the phase transformation of martensite into austenite distinguishes in temperature from structural recrystallization. Structural heredity refers to the reproduction of the initial austenitic grains in shape, size, and orientation upon heating. Figure 7.40 exemplifies how repeated austenitization restores the microstructure and orientation of an austenite grain.

The austenite grain restored retains (at least partially) the substructure created by the martensitic transformation. Otherwise speaking, the grain rests in a state of phase hardening. At a higher heating temperature, the restored grain suffers recrystallization, which results in the refinement of an initially large austenitic grain. Thus, the heating dictates the transformations to be a two-stage recrystallization scheme: first, phase transformation, then recrystallization (V.D. Sadovsky and K.A. Malyshev). Figure 7.41 presents a two-stage recrystallization scheme using high-alloyed steel as an example: an initial grain (a), a restored grain (b), recrystallization process (c), and its completion (d).

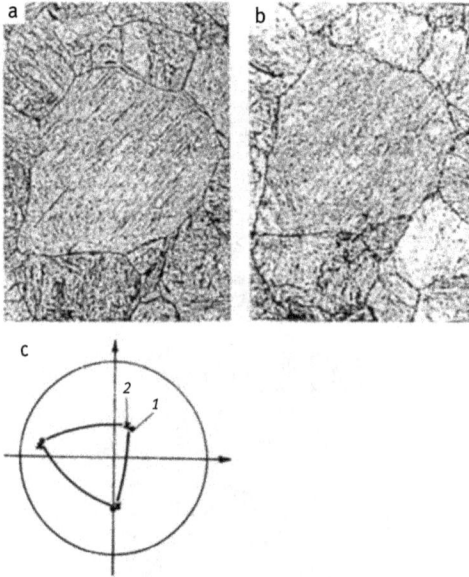

Figure 7.40: Recovery of a grain in medium-carbon alloyed steel: (a) an initial grain ($\times$ 200); (b) after requenching ($\times$ 200); and (c) grain orientation before (1) and after (2) reheating.

The manifestation of structural heredity depends on the degree of alloying of steel and the heating rate (Table 7.2). High-alloyed steels exhibit structural heredity at all heating rates (sign "+"); low-alloyed steels exhibit structural heredity only at very high heating rates.

A diagram of the influence of the heating rate on the grain's structural heredity is shown in Figure 7.42 for carbon and alloyed steels.

The non-monotonic influence of the heating rate for alloyed steels has been explained either by a change in the mechanism of formation of austenite with a change in the heating rate (V.D. Sadovsky and B.K. Sokolov) or by the intervention of recrystallization in the process of oriented (crystallographically ordered) transformation (V.D. Sadovsky and S.S. Dyachenko). With an increase in the heating rate, the degree of overheating above the phase equilibrium temperature increases. Consequently, the driving force of the transformation increases too. This can cause a change in the transformation mechanism. Using the example of low-carbon iron–nickel alloys (in the absence of carbides), it is established that during slow heating, the formation of austenite occurs orientedly, with the varying concentrations of substitutional and interstitial atoms (nickel and carbon). Figure 7.43 offers an illustration of oriented formation of austenite during slow heating of an alloy with the initial structure of equiaxial ferrite.

The concentration of nickel in grown-up Widmanstätten austenite crystals is increased compared to that in the initial ferrite. The experimental data obtained make it

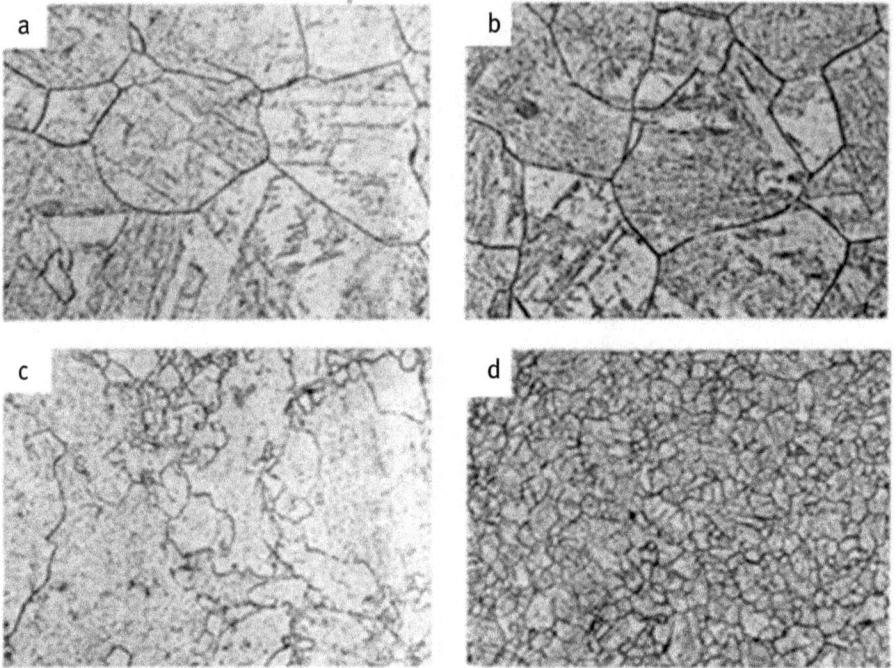

Figure 7.41: Two-stage recrystallization process in martensitic aging Cr–Ni steel. (×100): (a) an initial grain; (b) a recovered grain; (c) recrystallization process; and (d) finish of recrystallization with grain reduction.

Table 7.2: Dependence of structural heredity on the heating rate.

Steels	Quick heating, 1,000 deg/min	Moderate heating, 5–500 deg/min	Slow heating, 1–2 deg/min
High-alloyed steels	+	+	+
Alloyed steels	+	−	+
Low-alloyed and carbon steels	+	−	−

possible to classify the transformation as the decomposition of an α-solid solution with the precipitation of a nickel-rich γ-solid solution. With an increase in the heating rate, no diffusion redistribution of nickel occurs, and the transformation is brought about non-orientedly, similarly to the massive growth of a new phase. Massive growth is carried out by the fluctuation attachment of atoms or their small groups to a growing crystal. A sufficiently high heating rate without diffusion and relaxation processes results in the formation of austenite through a shear martensite-like mechanism.

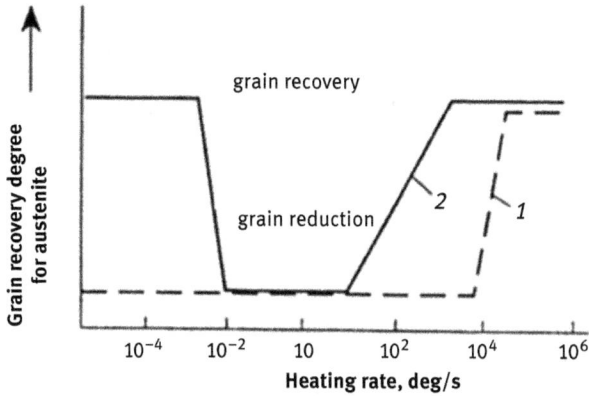

Figure 7.42: Heating rate influence on the grain recovery effect: 1, carbon steels; 2, alloyed steels.

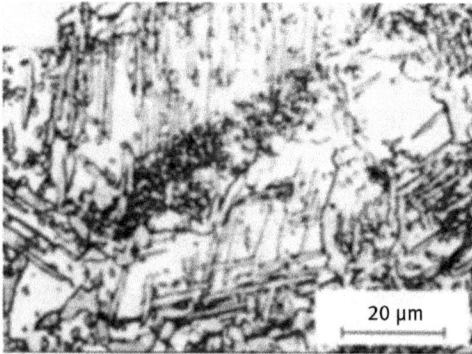

Figure 7.43: Oriented formation of austenite in a low-carbon Fe–Ni alloy with the initial structure of equiaxial ferrite. Slow heating at a rate of 0.05 deg/min.

Classification signs of the transformation of ferrite (martensite) into austenite are listed in Table 7.3.

Table 7.3: Characteristics of transformation of ferrite (martensite) into austenite in low-carbon iron–nickel alloys.

Transformation type	Nucleation place	Growth mechanism	Shape of crystals	Redistribution of components		Orientation relationship of phases
				Ni	C	
Decomposition	Volume of grains	Fluctuation	Rods and lamellae	Yes	Yes	Yes
Massive	Grain boundaries	Fluctuation	Globules	No	Yes	No
Martensitic	?	Shear	Lamellae	No	?	Yes

In the event of breaking the orientation relationship of the phases, there is no structural heredity. In this case, structural recrystallization is combined with phase transformation. The oriented growth of austenite in steel with an initial martensitic structure triggers a question: why does one of all possible crystallographically equivalent orientations of austenite get preferred and coincide with the initial one? When rapidly heated, the martensitic structure selects the favorable orientation because of its features, which preserved prior to the transformation. Among them are, firstly, retained austenite lamellae located between laths of martensite. They serve as ready-made nuclei, and the epitaxial growth on them determines the uniqueness of the orientation of austenite in each initial grain. Another peculiarity is a dislocation structure and internal stresses. The latter dictate the reversibility of atomic displacements with respect to those that occurred during the martensitic transformation. This situation is similar to the thermoelastic martensitic transformation. When slowly heated, austenite arises predominantly at the boundaries of martensitic crystal laths. In the process, the uniqueness of its orientation is governed by the simultaneous orientation relationship with neighboring martensite laths (V.M. Schastlivtsev). Figure 7.44 depicts electron microscopic images and a scheme of such a case of nucleation.

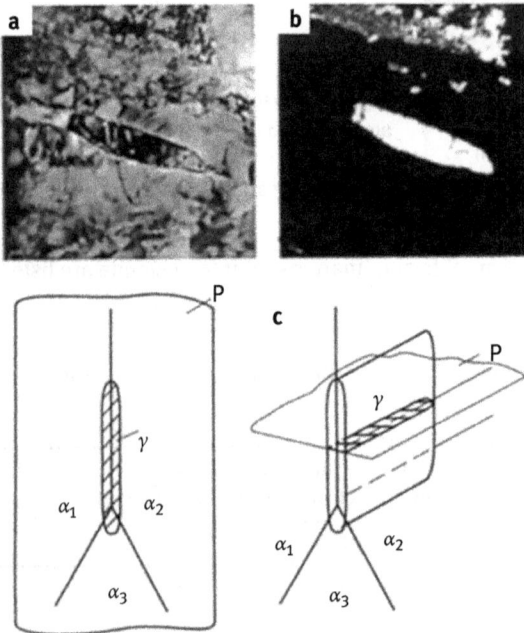

Figure 7.44: Nucleation of austenite at the boundary of laths: (a) structure ($\times 56,000$), (b) dark-field image in the reflection of austenite ($\times 56,000$), and (c) scheme of austenite nucleation at the boundary of tree laths of α-phase. P, plane of a foil.

Once combined, identically oriented grown-up nuclei create a structure with an initial orientation in each initial austenite grain.

Breakage of structural heredity during rapid heating is possible to take place due to the "grain boundary effect" (V.D. Sadovsky). The latter exhibits itself as the emergence of new fine grains along the boundaries of the initial grains during heating (Figure 7.45).

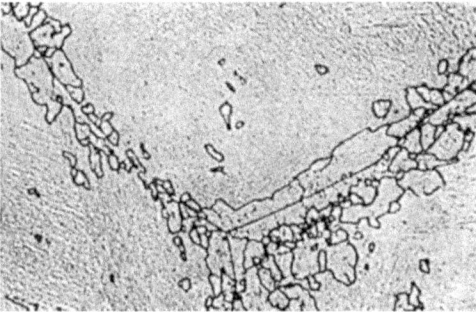

Figure 7.45: Formation of new grains of austenite along the initial boundaries ($\times 250$).

Distortions of the martensitic structure near the boundaries of the initial grains make it difficult for the austenite formation by the shear mechanism, and the mechanism of massive transformation begins operating. During slow heating, the structural heredity can be violated at the final stage of austenitization due to the formation of "white fields" (V.D. Sadovsky). In this case, the restored orientation austenite and still non-converted ferrite sites provoke the formation of completely converted austenitic regions (fields). Meanwhile, these ever-expanding fields capture reduced austenite and non-converted ferrite (Figure 7.46).

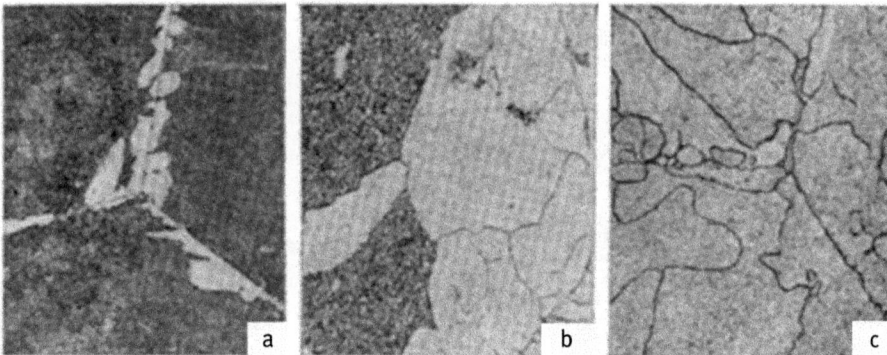

Figure 7.46: Growth of austenitic sites (white fields) at the final transformation stage in Cr–Ni steel ($\times 300$).

The pattern of the spread of austenitic fields in the austenite–ferrite structure (Figure 7.47) resembles a process opposite to that which occurs during discontinuous decomposition.

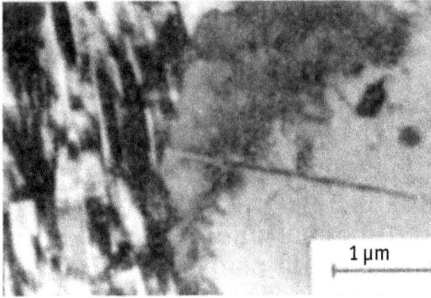

Figure 7.47: Absorption of lath (γ + α) structure by austenitic field. A Fe–Ni alloy. Heating rate of 0.4 deg/min.

The growth centers of the fields are austenite recrystallization nuclei. This process can be regarded as a recrystallization phenomenon combined with a phase transformation to reach equilibrium through diffusion along the migrating boundary. Alloying of steel with strong carbide-forming elements (e.g., titanium or vanadium) delays the recrystallization due to the precipitation of dispersed carbides without the growth of austenitic field centers. Thus, structural heredity in steel can be violated either by the massive (non-oriented) mechanism of the formation of new austenite grains at high and medium heating rates or by the recrystallization process of growth of new grains during slow heating at the final stage of transformation.

Conclusion

The book discusses in detail phase equilibria: numerous double phase diagrams and basic phase diagrams of ternary systems. Perceptions about temperature–pressure diagrams and metastable diagrams are given. The processes of crystallization and related phenomena, diffusion and its mechanisms, plastic deformation, including large deformations, polygonization, and recrystallization are considered. A large section is devoted to phase transformations in the solid state (martensitic transformations, decomposition of supersaturated solid solutions, and ordering). A separate chapter deals with the diverse transformations in steels and iron alloys. Almost all the sections exemplify transformations and structures of real alloys.

The main views on these issues are concisely and exactly summarized. Some interpretations of the phenomena are still controversial; in this case, the most common point of view is expressed. The recommended literature addresses the issues raised in detail.

https://doi.org/10.1515/9783110758023-008

Bibliography

Blank V.D., Estrin E.I. Phase Transitions in Solids Under High Pressure. CRC Press Taylor and Francis Group. Boca Raton. London-New York. 2014. 437.

Development of Ideas V.D. Sadovsky, A Member of the Russian Academy of Science: Collection of Treatises by V.M. Schastlivtsev, V.I. Zeldovich, D.A. Mirzaev, etc. Yekaterinburg, 2008. 410. [in Russian].

Gundyrev V., Zel'dovich V. About the mechanism of deformation at martensite transformation in the Fe–31 wt % Ni alloy. Mater. Sci. Forum. 2013. 738–739. 20–24.

Haidemenopoulos G.N. Physical Metallurgy: Principles and Design. CRC Press – Taylor and Francis Group. 2018. 490.

Hansen M., Anderko K. Constitution of Binary Alloys. 2nd Edition. McGraw-Hill Companies, Inc., New York. 1958.

Honeycombe R.W.K. The Plastic Deformation of Metals. Edward Arnold. London. 2nd Edition. 1984. 483.

Krivoglaz M.A., Smirnov A.A. The Theory of Order-Disorder in Alloys. Translated from the Russian edition (Moscow, 1958) by Scripta Technica Bruce Chalmers, 427.

Kurdyumov G.V., Utevskii L.M., Entin R.I. Transformations of Iron in Steel. Moscow. Nauka Publ., 1977. 238. [in Russian].

Physical Mesomechanics and Computer-Aided Design of Materials. Edited. by V.E. Panin. Novosibirsk. Nauka Publ. 1995. Vol.1. 298 Vol.2. 320 p. [in Russian].

Physical Metallurgy//Ed. by R.W. Cahn and P. Haazen. V.1. Atomic Structure of Metals and Alloys. North-Holland Physics Pub. 1983.

Physical Metallurgy//Ed. by R.W. Cahn and P. Haazen. V.2. Phase Transformations in Metals and Alloys and Alloys with Special Physical Properties. North-Holland Physics Pub. 1983.

Physical Metallurgy//Ed. by R.W. Cahn and P. Haazen. V.3. Physical and Mechanical Properties of Metals and Alloys. North-Holland Physics Pub. 1983.

Physical Metallurgy. 5th Edition//Editors: David Laughlin and Kazuhiro Hono. Elsevier. 2014. 2920.

Porter D.A., Esterling K.E. and Sherif M.Y. Phase Transformations in Metals and Alloys. Third Edition. CRC Press. 2015. 538.

Rybin V.V. Large Plastic Deformation and Fracture of Metals. Moscow. Metallurgiya Publ., 1986. 224. [in Russian].

Schastlivtsev V.M., Kaletina Yu.V., Fokina E.A. Martensitic Transformation in Magnetic Field. Yekaterinburg: Ur Br RAS, 2007. 322. [in Russian].

Shewmon P.G. Diffusion in Solids. McGraw-Hill, 1963, 238.

Smallman R.E. and Ngan A.H.W. Modern Physical Metallurgy. 8th Edition. Elsevier Ltd. 2014. 697.

Valiev R.Z., Alexandrov I.V. Bulk Nanostructured Metallic Materials: Obtain, Structure, Properties. Moscow. IKC Akademkniga. 2007. 398. [in Russian].

Zehetbauer M.J., Valiev R.Z. Nanomaterials by Severe Plastic Deformation. Weinheim, Germany: Wiley-VCH. 2004. 875.

https://doi.org/10.1515/9783110758023-009

Index

Abrikosov A.A. 8
activation energy 47, 66–67, 79–81, 107–108,
    130, 132, 168, 195
alloyed steel 163–165, 169, 171, 174, 176,
    181–182, 184, 188, 190–191, 198–200
amorphous alloy 7, 59–62
amorphous state 7, 53, 59–61, 63
Anosov P.P. 2
antiphase boundary 160
athermal kinetics 172
athermal transformation 131–132, 139
atomic ordering 156, 159
austenite VII, 2–3, 84, 127–130, 132–136,
    140–142, 153–154, 156, 163–173, 175–176,
    179–182, 184–188, 190–191, 195–199,
    201–204
austenite formation VII, 191, 203
austenite nucleus 196
austenitic 127
autocatalytic transformation 173
Avrami M. 6, 113

Bain E.C. 3–4, 134–136, 138
bainite 165, 182–188, 191
bainitic (intermediate) transformation 166
bainitic transformation VII, 165–166, 182–184,
    186–187
Bardeen J. 8
Bednorz J.G. 8
Bochvar A.A. 6, 8, 55, 58, 60, 115
boundary diffusion 79–80, 155, 170
Bozorth R.M. 8
Brownian motion 71
bulk diffusion 79–80, 170
butterfly-like martensite 176

Cahn R.W. IX, 7
cementite 156, 163–167, 169–172, 182, 184,
    187–192, 196–198
chemical potential 5, 11, 65, 82, 84–86
Chernov D.K. 1–2, 54–55
coarse-lamellar pearlite 171
coherent interphase boundary 151
cold plastic deformation 90, 96
collective recrystallization 111, 118–119

columnar crystals 55–56
conoid 17–18
continuous decomposition 146, 148, 150,
    153, 155
Cooper L. 8
Cottrell H.H. 6
critical nucleus 45, 134
crystallization VII, IX, 2–3, 5, 9, 15, 17, 22–23,
    25, 31–34, 36, 43–53, 56–63, 103,
    112–113, 125, 132, 205
Curie ordering temperature 156, 158

Darken experiment 65
decomposition of martensite 188, 191–192
decomposition of retained austenite 188,
    191, 193
decomposition of supersaturated solid
    solution VII, IX, 7, 9, 65, 125–126, 205
deformation texture 103, 120
degree of freedom 17
degree of order 4
degree of overcooling 43
degree of tetragonality 174
dendritic growth 51–52, 55
diffusion VII, IX, 2, 4–5, 9, 20, 46, 57–59, 61,
    65–87, 89–90, 108, 116–117, 125,
    127–128, 132, 144, 146, 150, 152–156, 158,
    167, 169–171, 183–185, 187, 191, 196–197,
    200, 204–205
diffusion coefficient 5, 46, 65–68, 70–72,
    78–79, 81–84, 87, 108, 154–155
diffusion in metals IX, 9
diffusionless 171
discontinuous decomposition 147, 154–156,
    167, 170, 204
dynamic recrystallization 120–121

elastic deformation 70, 89, 125, 140–142, 172
elastic strain energy 125
equal-channel angular pressing 101
equilibrium 4, 11–12, 14–15, 17–18, 20–23, 26,
    32, 34–37, 39, 43, 54, 57, 60, 70, 77, 84,
    86, 100–101, 118, 127, 129, 132, 145,
    147–148, 150–151, 153, 163, 166–167, 181,
    187, 195–196, 199, 204

https://doi.org/10.1515/9783110758023-010

Also of interest

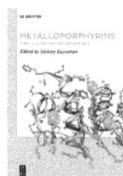

Metalloporphyrins.
Tuning Properties and Applications
Aleksey Kuznetsov (Ed.), 2022
ISBN 978-3-11-061747-4, e-ISBN 978-3-11-061802-0

Bulk Metallic Glasses and Their Composites.
Additive Manufacturing and Modeling and Simulation
Muhammad Musaddique Ali Rafique, 2021
ISBN 978-3-11-074721-8, e-ISBN 978-3-11-074723-2

Homogenisierungsmethoden.
Effektive Eigenschaften von Kompositen
Rainer Glüge, 2021
ISBN 978-3-11-071948-2, e-ISBN 978-3-11-071949-9

Metallschäume.
Herstellung, Eigenschaften, Potenziale und
Forschungsansätze – mit Schwerpunkt auf Aluminiumschäume
Günther Lange, 2022
ISBN 978-3-11-068155-0, e-ISBN 978-3-11-068175-8

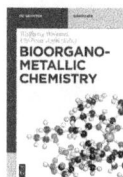

Bioorganometallic Chemistry
Wolfgang Weigand, Ulf-Peter Apfel (Eds.), 2020
ISBN 978-3-11-049650-5, e-ISBN 978-3-11-049657-4

www.ingramcontent.com/pod-product-compliance
Lightning Source LLC
Chambersburg PA
CBHW061414210326

41598CB00035B/6209